For the
Origin of
Free Energy

OTHER BOOKS BY Makoto D. Yamane

Essay "Window" the 17th collection "Resonance frequency," Makoto D. Yamane, MEISOU Publishing Inc. Dec.20th, 2010 P184

For the Origin of
Free Energy

New "Power and Energy Theory"

By Energy • Volunteer

Makoto D. Yamane

MEISO CANADA PUBLISHERS

Printed in the United States of America

First edition: 2012

Limit of Liability and Disclaimer of Warranty
The publisher and the Author of this book have used their best efforts in preparation of this book and make no representations or warranties of any kind, expressed or implied, with respect to the accuracy or completeness of the contents of this book. The fact that an organization or website is referred to in this work as a citation and/or a potential source of further information does not mean that the author or the publisher endorses the information the organization or website may provide or recommendation it may make. Readers should be aware that any internet website listed in this work may have changed or disappeared at the time when it is read.

ISBN 978-0-9878009-0-9

DEDICATION

This book is dedicated to my late father, my
late mother, and my wife Keiko.

ACKNOWLEDGEMENTS

I wish to express my gratitude to these persons:

Alexander V. Frolov------- "Free Energy & Antigravity Technology" Russia

Chris Sykes----------------- "SQM Replication Project" USA

Ismael Aviso---------------- "Self Charging Electric Car" Philippines

Jeane Manning------------ "Change Power Net" Canada

Jerry Decker---------------- "Keely Net" USA

Mark Hendershot---------- "Mark III" USA
Figure 4.3.1, 4.3.2

Paul E. Potter--------------- "M-L Converter" England
Figure 4.5.1 – 4.5.4

Sterling Allan---------------- "PESWiki" USA

Thorsten Ludwig----------- "Hans Coler" Germany
Figure 3.6, 3.7, 4.2.1, 4.2.2-1, 4.2.2-2

Tim Harwood and John Jankowski
----------- "POD II" USA
Figure 3.1.1, 4.4.1 – 4.4.3

Violet Sweet----------------- "SQM" USA
Figure 4.6.1 – 4.6.3

Wang Shum Ho------------ "Self Running Generator" China

William S. Alek------------- "Newtonian Torsion Physics" USA
Figure 3.3

Table of Contents

Preface

I went to Makoto D. Yamane's publishing ceremony where he presented a pretty clear explanation of free energy. However, in this book Makoto D. Yamane offers a more extensive explanation.

The book provides us with an almost complete understanding of free energy, and has been recognized as an important resource on this topic.

This brilliant book impresses even engineers, and it gives them hope. One of the very helpful resources is the attached diagram in Chapter 4.

What is the root of free energy? You can find a clear answer to the above question when you read this superb book!

I believe all matter in the universe originated from the photon, and all the energy used in the earth's composition started from "Photon" at the Earth in its early stage.

Understanding the details of free energy is not easy for me, but the description of the free energy of Nikola Tesla's coil and the variety of Tesla's inventions are very evocative.

The author's reference to Prof. Lisa Randall's work is indeed enlightening.

Thanks for this great source of knowledge! I know that Love is Free Energy. Thank you all!

Yuka Nakagawa

Introduction

I am not a physical researcher, but I like Thomas Edison very much and worked as an engineer in the three electric companies.

The spread of free energy devices is my lifework.

The conclusion of my researching the free energy circuits gave me a chance to write this book.

I think that I am lucky to have approached directly to "Power," not "Energy."

This book consists of Chapter 1 – Chapter 4.

Chapter 1 and Chapter 2 are the chapters of the theories.

Chapter 3 describes elemental technologies of the free energy devices, and Chapter 4 is a technological chapter of the circuits with the free energy.

I would like you to confirm the real images and the principles of the elemental technologies and the circuits with the free energy devices.

I am very honored that I am able to open the theory and the technology which will be a foothold to solve the problems in the future.

You can use this book as a guide to the free energy and its devices.

February, 23, 2011

Energy • Volunteer
Makoto D. Yamane

Chapter 1

For the Origin of Free Energy

1.1 New "Power and Energy Theory"

In "Alice in Wonderland," which Lewis Carroll wrote, Alice jumps into a hole in pursuit of a white rabbit and falls into her dream in an underground world. She tries to run after the rabbit and encounters many misfortunes.

"Warped Passages"[1] by Lisa Randall is a masterpiece where the abstruse theory is summarized plainly and concisely. When I am uncertain about its content I see it as a guidepost that pops up similarly to Alice's rabbit, and I use it despite not being certain of its meaning even though I continue to read.

There is a mythical story called "White rabbit in Inaba" from my hometown, Tottori City. When the rabbit has been skinned and is suffering, Okuninushinomikoto passes by and saves her. Later the rabbit arranges a marriage between Okuninushinomikoto and Yakamihime.

I employ these two rabbits, the matchmaking rabbit and the guiding rabbit, to solve the problems of power and energy for the origin of free energy.

There are four kinds of power—Electromagnetism Power, Gravity, Strong Power, and Weak Force.
    An example of Strong Power is manifested when quarks unite in the proton and the neutron.
    An example of Weak Force is manifested when the neutron decays the beta ray and changes into the proton.
    Electromagnetism Power and Weak Force were unified by the "Electroweak Theory" in the 1960's.
This leads us to think about how Electromagnetism Power and

Gravity are related. Further, what is Strong Power? I found that it is faster to master the fundamentals of the integrated powers as a result of slipping down various holes.

Weak Force relates to the reaction of neutron → proton +   Weak Gauge Boson, Weak Gauge Boson   →   electron + anti-electronic type neutrino. There is a problem here: the Weak Force breaks the parity symmetry. It has been experimentally proven that only a counterclockwise particle can receive the action of the Weak Force. However, the reason is not explained in terms of intuition and dynamic knowledge. Moreover, there are a lot of uncertain points that Weak Gauge Boson has mass and spin one.

Recently I reread the book written by Shuji Inomata, "Paradigm of New Science: Principia for the 21st Century." [2] The reason for this is as follows: In his book, I read "Secret of Mr. Kimura's Apple Miracle[3]" written by Yasuhisa Oharada. He noticed the fact that Akinori Kimura developed the apples with no agricultural chemicals.

Mr. Kimura's story was as follows: On a certain rainy day Kimura went to buy a book on turbocharged engines. When he accidentally knocked a book down from the top shelf with his stick, it happened to be a book he didn't want. The book became wet so he bought that book without seeing its title. The book was left covered in dust without being read. One day Kimura casually glanced at the book and noticed the title, "Natural Farming" written by Masanobu Fukuoka. The content of the book was a great shock to him, and Kimura read the book repeatedly.

I have a similar experience to that of Mr. Kimura. In my case the book is the "Paradigm of New Science: Principia for the 21st Century."
    At the time, in 1995, I was coming and going on overseas trips to Hong Kong and China as a factory adviser. One Sunday when reading a newspaper from Japan, I found an introductory article

about "Paradigm of New Science: Principia for the 21st Century." A month was necessary to find this book via special request from my mother in Japan. Strolling through the park was my Sunday habit. Same Sunday I happened to go into a bookstore in Hong Kong's Daimaru Department store, and I found something glittering among the cookbooks. When I picked it up, the book was "Paradigm of New Science: Principia for the 21st Century." Needless to say I was astonished. I read the book more than ten times.

This book is about the developing of Maxwell's electromagnetic equation by the complex number. After rereading it, I found that the triangular converting formula chart of Consciousness (Q), Matter (M), and Energy (E) remains the masterpiece of this book. It is thought that Q of consciousness is equivalent to charge Q of the Electromagnetism Power.

Power F shows "F= G × M1•M2/r^2,
G: Newton's gravitational constant"
and "F= Kc × Q1•Q2/ r^2, Kc: Coulomb's constant."

1.2 Strong Power

It is easy to understand the Aether Physics Model[4] quantifying quantum structure including the unit system. I pursue the subject by using the Aether Physics Model by which Strong Power is made aether. When you put the quantum measurements of the mass and the strong charge in the following expression we see that:

Proton	+	Electron
1.673×E-27 kg		9.109×E-31 kg
2.57×E-34 $coul^2$		1.4×E-37 $coul^2$

Neutron (Proton, Weak Force One) + →(Strong Power)←	
1.675×E-27 kg	3.268×E15 kg
2.573×E-34 coul²	5.021× E 8 coul²

I calculate the "Weak Force One" mass and the strong charge from the value of the neutron and the proton. The Weak Force One mass and the strong charge are about twice the electron mass and the strong charge. Weak Force One becomes:

21.8×E-31 kg (about twice the electron mass)
3.35×E-37 coul²

In the same way:

Neutron (Proton, Weak Force One) + →(Strong Power)←	
1.675×E-27 kg	3.268×E15 kg
2.573×E-34 coul²	5.021× E 8 coul²

Proton + Weak Force One + ←(Weak Force Two, Positron)→		
1.673×E-27 kg	21.8×E-31 kg	-12.691×E-31 kg
2.57×E-34 coul²	3.35×E-37 coul²	-1.95× E -37 coul²

Proton +	Electron +	Electron +	Positron +	Neutrino
1.673×E-27 kg	9.109×E-31 kg	9.109×E-31 kg	-9.109×E-31 kg	
2.57×E-34 coul^2	1.4×E-37 coul^2	1.4×E-37 coul^2	-1.4×E-37 coul^2	

Proton +	Electron +	Photon +	Neutrino
1.673×E-27 kg	9.109×E-31 kg	0 kg	
2.57×E-34 coul^2	1.4×E-37 coul^2	0 coul^2	

This shows as following:

Photon =	Electron +	Positron
0 kg	9.109×E-31 kg	-9.109×E-31 kg
0 coul^2	1.4×E-37 coul^2	-1.4×E-37 coul^2

   "←(Weak Force Two)→" in the expression above the mass and the strong charge are negative.

The name of the book, "Blind Spot of the Contemporary Physics" by Kenichi Konno, which I read 30 years ago, was changed to "True Colors of Super-Trick Dark Matter of the Cosmology"[5] on September 30, 2010. At the time it was published as a revised edition.

According to this book the positron is equal to ←(Weak Force Two)→. In addition, the physical properties of the space (vacuum), which was pointed out by Kenichi Konno, are the arrows of ←(Strong Power)→. Kenichi Konno further explained,

"Aether is an origin where the material is made." When the ratio of the strong charge to the mass for each within the Aether Physics Model is calculated for each of following, it can be seen $6.51×E6$ kg/coul^2 and it is constant[6]:

• Electron $9.109×E{-31}$ kg / $1.4×E{-37}$ coul^2= $6.5064×E6$ kg/ coul^2

• Proton $1.673×E{-27}$ kg / $2.57×E{-34}$ coul^2= $6.5097×E6$ kg/ coul^2

• Neutron $1.675×E{-27}$ kg / $2.573×E{-34}$ coul^2= $6.5099×E6$ kg/ coul^2

•→(Strong Power)← $3.268×E15$ kg / $5.021×E8$ coul^2= $6.5087×E6$ kg/ coul^2 (I add that the physical properties of the space is arrow.)

• Weak Force One $21.8×E{-31}$ kg / $3.35×E{-37}$ coul^2= $6.5075×E6$ kg/ coul^2 (I add that Weak Force One mass is about equal to the mass of two electrons.)

• ←(Weak Force Two, Positron)→ $-12.691×E{-31}$ kg / $-1.95×E{-37}$ coul^2= $6.5082×E6$ kg/ coul^2 (I add that the physical properties of the space is arrow and Weak Force Two is positron.)

• Positron $-9.109×E{-31}$ kg / $-1.4×E{-37}$ coul^2= $6.5064×E6$ kg/ coul^2

1.3 Photon

I will consider the photon in this section.

"Let there be light!" is a phrase that is found in Genesis. I had never thought about how light had such a huge task. However, if we think about things such as the light of the sun, we have always been extremely indebted to light.

The following is understood about the photon:
• The photon has zero mass and charge, and it is a steady elementary particle.
• It is a gauge element that mediates an electromagnetic interaction.

On the other hand, the photon is shown by $hc= 16\pi^2\, Kc\, eemax^2$ in the Aether Physics Model. When we assume the mass of →(Strong Power)← to be ms, and the charge of →(Strong Power)← to be es is generated with the Casimir effect. This is due to the charge which the electron receives the action according to power with a strong virtual photon.

(1) $G \bullet ms^2 / \lambda c^2 = $ G force   G: Newton's gravitational constant

λc: Compton wavelength

(2) $16\pi^2 \bullet Kc \bullet es^2 / \lambda c^2 = $ G force   Kc: Coulomb's constant

h: Planck's constant

me: Electronic mass $eemax^2$: Strong charge of electron

$me= h / (c \bullet \lambda c)$, $ms / es^2 = me / eemax^2$

(3) $ms / es^2 = (\sqrt{Gforce} / \sqrt{G}) \bullet \lambda c \bullet 16\pi^2 \bullet Kc / (Gforce \bullet \lambda c^2)$

$= 1 / (\sqrt{Gforce} \bullet \sqrt{G} \bullet \lambda c) \bullet hc / eemax^2$

$$= 1 / (\sqrt{G} \cdot \sqrt{Gforce}) \cdot (m_e \cdot c^2) / e_{emax}^2$$

$$= 1 / (\sqrt{G} \cdot \sqrt{Gforce}) \cdot c^2 \cdot (m_e / e_{emax}^2)$$

$$= m_e / e_{emax}^2$$

(4) $\sqrt{G} \cdot \sqrt{Gforce} = c^2$ c: Speed of light

$$= \sqrt{G^2} \cdot \sqrt{m_s^2} / \sqrt{\lambda_c^2} = G \cdot m_s / \lambda_c$$

(5) $m_s = c^2 \cdot \lambda_c / G$ $m_e = h / (c \cdot \lambda_c)$

$$= c^2 \cdot h / (c \cdot m_e) \quad G = hc / (m_e \cdot G)$$

• **Photon** $= hc = 16\pi^2 Kc$ $e_{emax}^2 = G \cdot m_e \cdot m_s$

• **The mass m_s of →(Strong Power)← is taken into the photon**.

1.4 True Colors of Dark Matter and Dark Energy

Because consideration of the photon and →(Strong Power)← mass is taken into the photon, the charge of Strong Power remains:

• es^2= ms • $eemax^2$/ me= $(c^2 • λc) / G • hc/ 16π^2 Kc • (c • λc) / h$

$$= c^4 • λc^2 / (G • 16π^2 Kc)$$

It is understood that it is Dark Energy which forms charge es^2 of this →(Strong Power)←, and this Dark Energy is the origin of free energy.

As for the aforementioned Dark Energy, in the past, the theory that stated its relation to Aether ($16π^2Kc$) and Gravity (G) was not wrong, due to the content of the composition with the expression that it contains both aether and gravity.

It is evident that the Dark Matter is ms= $c^2 • λc / G$ and mass of →(Strong Power)← from the understanding of Dark Energy.

The Gauge Boson of the Weak Force becomes Weak Force One + ←(Weak Force Two) →. Strong Power diffuse becomes a positron with the former mass and the charge. Therefore, correspondingly it becomes only a counterclockwise particle because it has clockwise motion. Finally, it returns to electron (e-) by Weak Force One (2e-), and ←(Positron e+ Weak Force Two)→ generates the photon. There is mass as well, and the spin becomes one. In addition, it is thought that the neutrino is caused in that case.

The New Weak Gauge Boson Hypothesis is discovered thanks to Alice's rabbit. It is thought that the traditional way of thinking does not overlook, but also does not know where mass and the charge of Strong Power went because we paid too much attention to the neutron.

1.5 New Interpretation of Gravity "Breaking Down of Hierarchy"

According to conversion formula of Consciousness • Charge (Q), Matter (M), and Energy (E) "Paradigm of New Science: Principia for the 21st Century," it becomes $G = (Q/M)^2$. When I calculate using the Aether Physics Model:

$G = 16\pi^2 \cdot Kc \, (es / ms)^2 = hc \, (photon) / (me \cdot ms)$

It follows from this that "Gravity is power to work between the mass of electron and Strong Power." The reason is that it can explain Planck's constant h and the hierarchy with mass of electron by this formula. The hierarchy has been broken down. I owe this completely to Alice's rabbit as well.

1.6 Integrated Hypothesis of Four Fundamental Forces

"Electromagnetism Power," "Gravity," "Strong Power," and "Weak Force"

The previous expression 1.2 is brought together as follows:

Proton	+	Electron	+	Photon	+Dark Matter, Dark Energy
1.673×E-27 kg		9.109×E-31 kg		0 kg	
2.57×E-34 coul^2		1.4×E-37 coul^2		0 coul^2	

Proton	+	Electron	+	Electron (Photon)	+ Positron (Photon)
1.673×E-27 kg		9.109×E-31 kg		12.691×E-31 kg	-12.691×E-31 kg
2.57×E-34 coul^2		1.4×E-37 coul^2		1.95×E-37 coul^2	-1.95×E-37 coul^2

+

Dark Matter, Dark Energy

Proton	+	Weak Force One +	→(Strong Power)←
1.673×E-27 kg		21.8×E-31 kg	3.268×E15 kg
2.57×E-34 coul^2		3.35×E-37 coul^2	5.021×E8 coul^2

Neutron	+	→(Strong Power)←
1.675×E-27 kg		3.268×E15 kg
2.573×E-34 coul^2		5.021×E8 coul^2

Here:

> • Weak Force One= Electron+ Electron (photon)
>
> • →(Strong Power)← = →Positron(photon) ← + Dark Matter
>
>   (Arrow is physical properties of the space)   Dark Energy
>
> • Mass ms : hc (photon)= G • me • ms
>
>   ms= hc(photon) / G • me= Dark Matter
>
> • Charge es^2: It depends on the Casimir Effect.
>
> es^2= c^4 • λc^2 / (G • 16π^2 Kc)= Dark Energy

Neutron(Proton, Weak Force One)	+ →(Strong Power)←
1.675×E-27 kg	3.268×E15 kg
2.573×E-34 coul^2	5.021×E8 coul^2

Proton	+	Weak Force One +	←(Weak Force Two)→
1.673×E-27 kg		21.8×E-31 kg	-12.691×E-31 kg
2.57×E-34 coul^2		3.35×E-37 coul^2	-1.95×E-37 coul^2

Proton	+	Electron	+	Electron(Photon)
1.673×E-27 kg		9.109×E-31 kg		9.109×E-31 kg
2.57×E-34 coul^2		1.4×E-37 coul^2		1.4×E-37 coul^2

+

Positron(Photon)(Including ms) + Dark Energy (es^2) + Neutrino		
-9.109×E-31 kg	0 kg	a little kg
-1.4×E-37 coul^2	5.021×E8 coul^2	0 coul^2

⇩

Proton	+	Electron	+	Photon
1.673×E-27 kg		9.109×E-31 kg		0 kg
2.57×E-34 coul^2		1.4×E-37 coul^2		0 coul^2

+

Dark Energy	+	Neutrino
0 kg		a little kg
5.021×E8 coul^2		0 coul^2

• Weak Force Two= Positron (photon) + Neutrino

Weak Force Two becomes as ←(Strong Power)→ divergence.

> **"Electromagnetism Power"**   **Gauge: Photon**
>
> **"Gravity"** G= hc (photon) / (me• ms)   **Gauge: Photon**
>
> **"Strong Power"** Mass ms= hc(photon)/(me•G) **Gauge: Photon**
>
> Charge es^2= c^4 • λc^2 / (G• $16\pi^2$Kc)   Casimir Effect
>
> **"Weak Force"**= Weak Force One + Weak Force Two
>
> ="Electron+ Electron(photon)"+ ←Positron(photon)→
>
> **Gauge: Photon**

All four fundamental forces—"Electromagnetism Power," "Gravity," "Strong Power" and "Weak Force"—are related to the photon and they are united with the Gauge Photon.

An integrated hypothesis of new power is born, thanks to the rabbit of connection.

1.7 The Origin of Free Energy

The origin of free energy is charge es^2 of →(Strong Power)←, and is called Dark Energy. If it is possible to extract this by using physical properties of the space (vacuum) and Electromagnetism Power, we can also use it as energy.

1.8 The Universe

• Gravity is Power to work between the Mass of Electron and Strong Power.

• The Dark Matter is Mass of Strong Power.

• The Dark Energy is a Charge of Strong Power.

• The Mass of Strong Power is taken into the Photon.

• The Origin of Free Energy is a Charge of Strong Power and Dark Energy.

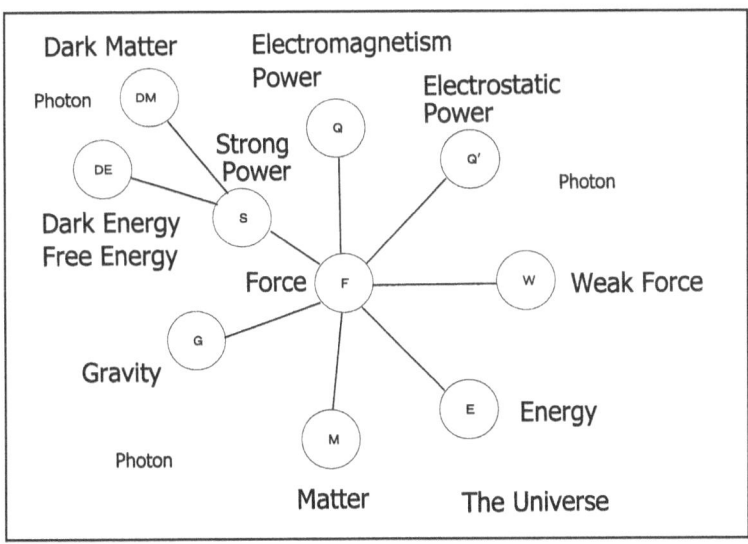

Figure 1.8.1 The Universe

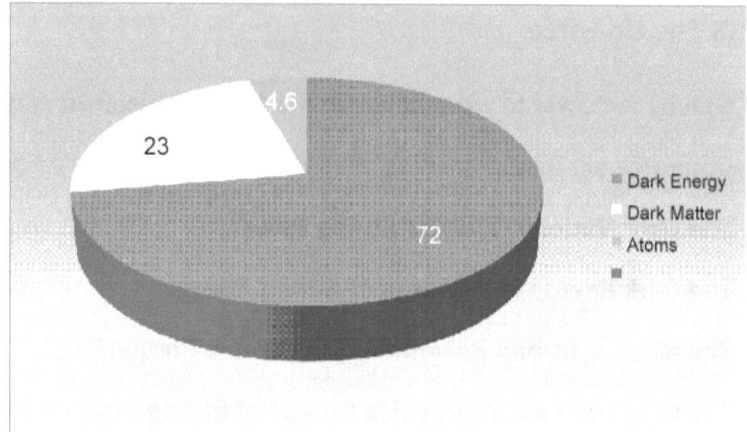

Figure 1.8.2 Estimated Distribution of Dark Matter and Dark Energy in the Universe

Dark Energy= Charge $es^2 = c^4 \cdot \lambda c^2 / (G \cdot 16\pi^2 Kc)$ = **Free Energy**

$es^2 = 5.021 \times 10^8$ coul^2 and ms/ $es^2 = 16\pi^2 Kc / c^2 \cdot \lambda c$

E $es^2 =$ **E** Free Energy= ms c^2 = $(16\pi^2 Kc / c^2 \cdot \lambda c)$ es^2 c^2

$= (16\pi^2 Kc / \lambda c)$ es^2

$= (1.419 \times 10^{12} / 2.426 \times 10^{-12}) \times 5.021 \times 10^8$

$= 1.833 \times 10^{51}$ eV$= 2.936 \times 10^{32}$ J=**2.9×10²⁹ kW**

=**2.9×100,000,000,000,000,000,000,000,000,000 kW**

Dark Matter= Mass ms= c$^2 \cdot \lambda c / G$= 3.268×10^{15} kg

1.9 Summary

• Photon= hc= $16\pi^2$Kc $e$emax^2= G • me• ms

   Mass ms of →(Strong Power)← is taken into the photon.

• The Dark Matter is ms= c^2 • λc/G= 3.268×10^{15} kg.

• es^2= c^4 • λc^2 / (G • $16\pi^2$Kc)= 5.021×10^8 coul^2.

The Dark Energy forms charge es^2, and this Dark Energy is

the origin of free energy.

• As for the Dark Energy the theory says this relates to Aether

($16\pi^2$ Kc), and Gravity (G).

• G= hc(Photon) / me • ms   Gravity is power to work between

the mass of electron, and Strong Power.

• All four Powers "Electromagnetism Power," "Gravity," "Strong

Power" and "Weak Force" are united with the Gauge Photon.

• If it is possible to extract this by using physical properties of

the Space (vacuum) and the Electromagnetism Power we can

use it as energy that is the charge es^2 of →(Strong Power)←

and is called Dark Energy.

• **E** es^2= **E** Dark Energy= **E** Free Energy= 2.93×10^{29} kW

1.10 Reference

1. "Warped Passages," Lisa Randall, Japan Broadcast Publishing Co., Ltd, June 30th, 2007

2. "Paradigm of New Science: Principia for the 21st Century," Shuji Inomata, Gijutu Shupan Co., Ltd, October 10th, 1987

3. "Secret of Mr. Kimura's Apple Miracle," Yasuhisa Oharada, Gaken Co., Ltd, March 10th, 2010

4. "Secrets of the Aether," David W. Thomson III and Jim D. Bourassa, Quantum Aether Dynamics Institute, Third Edition 2007

5. "True Colors of Super-Trick Dark Matter of the Cosmology," Kenichi Konno, Hikaru Land Co., Ltd, September 30th, 2010

6. http://oriharu.net/jAPM.htm
 "A New Foundation for Physics," Quantum Aether Dynamics Institute, Translated to Japanese by Mr. Oriharu

Chapter 2

Fundamental Equation to Obtain
Free Energy (Negative Current)

2.1 Maxwell's Electromagnetic Equation

Maxwell's electromagnetic equation is as follows:

2.1.1 $\nabla \cdot E = 4\pi\rho$ E: Electric Field

                                            c: Speed of Light

2.1.2 $\nabla \times E = -1/c \cdot \partial B/\partial t$ B: Magnetic Field

2.1.3 $\nabla \cdot B = 0$ ρ: Charge

2.1.4 $\nabla \times B = 4\pi \cdot J/c + 1/c \cdot \partial B/\partial t$ J: Current

This is an electromagnetic equation using a positive current, and it doesn't appear to show the free energy principle. For example, a given mA of input is increased to 1A.

2.2 Another Electromagnetic Equation of Maxwell

There was another variant of Maxwell's electromagnetic equation, but some of it was omitted in 1890 by Heinrich Hertz in the process of simplifying the equation.

2.2.1 $-E = \nabla \cdot \phi + \partial A/\partial t$ ϕ: Scalar Potential

2.2.2 $B = \nabla \times A$ A: Vector Potential

2.2.3 $\partial\rho/\partial t = \nabla \cdot J$

1. The scalar potential and the vector potential are generated by the coil as explained in the basic technology of the free energy device in Chapter 3.

2. $\nabla\times$indicates rotation(rot), $\nabla\bullet$ the divergence (div). They exhibit physical properties in a space (vacuum) that are explained in Chapter1. The equations are believed to be $\nabla\times= \rightarrow$(Strong Power)$\leftarrow$, and $\nabla\bullet = \leftarrow$(Strong Power)$\rightarrow$.

3. Equation 2.2.1 illustrates conditions that turn an electric field negative, and equation 2.2.2 illustrates turning a magnetic field positive and rotating it.

The procession around the magnetic field is similar to a toy spinning top, as illustrated in Figure 2.2, and is explained by Shinichi Seike in the Principle of Ultra Relativity "Tairiku Shobou." It is called axis rotation. Spinning around the electric field is called polarity rotation.

Axis rotation and polarity rotation are always perpendicular to each other. This shows the conditions of dropping the polarity rotation to the bottom half to suppress it to negative energy. This is believed to illustrate the basic requirements necessary to drive the free energy device.

4. Equation 2.2.3 shows the turning of the charge into an electric current.

Based on 1 through 4, Maxwell's equation using electrical magnetic potential is the fundamental equation for obtaining Free Energy "Negative Current."

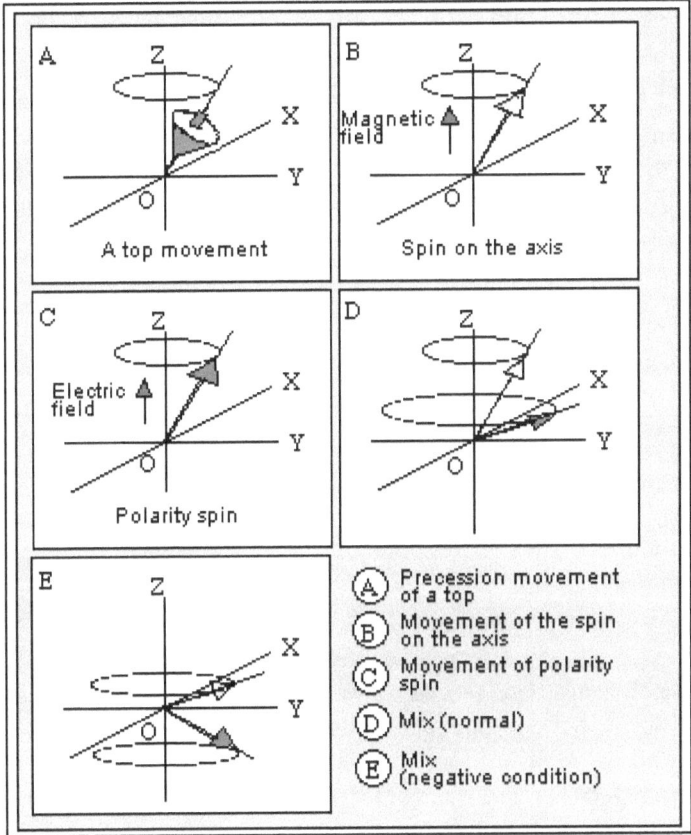

Figure2.2 A Spin[1]

2.3 Summary

• Maxwell's electromagnetic potential equation is fundamental to obtain Free Energy "Negative Current."

2.4 Reference

1. http://www.geocities.co.jp/Technopolis/1123/seike.html

2.5 Clipping Data

• In the original paper "Kinetic theory of the electromagnetic field" by Maxwell (1865) and original textbook "Electric magnetism theory" (1873), the electromagnetic potential 2.2.1, 2.2.2 were shown first, in place of expression 2.1.2 and 2.1.3.

• Regarding the scalar potential and the vector potential:

- The charge makes the scalar potential and the current makes the vector potential.

- Electromagnetic potential energy is decided by the vector potential and the scalar potential.

• As a positive current: there is generation of heat when energizing and there is resistance. As a negative current: Endothermic becomes cold when energized. It is the same as in the Orient, the opposite of negative is called positive. It is called a cold current because it becomes cold when energized. Complex current is the same current, which also became known as the cold current due to the cancellation by the Nikola Tesla invention, though this jumps ahead to Chapter 3.

Negative Current = Cold Current = Complex Current

Chapter 3

Elemental Technology of Free Energy Device

The following elemental technologies form the free energy device:

1. Use of Cold Current and Negative Area
      1. Method with Permanent Magnet
      2. Method of Using the Moebius Ring
2. Increasing Power with Coil
3. Use of Complex Field "Magnetic Field Cancellation with Coil"
4. Use of Back Electromotive Force with Coil
5. Use of Pulse
6. Circuit for Making to Current of Electromagnetic Charge
7. Output Circuit
8. Resonant Circuit
9. Oscillation Circuit

These are the nine elemental technologies—a collection of common technologies invented by our predecessors. I don't believe that all these technologies are necessary. I think that we will be able to design a new device by combining these technologies.

There is a super-efficient device that improves the power generation efficiency of the positive energy. Heat is generated in the process of energizing from resistance. When I energize, I introduce the elemental technology that is necessary for the free energy device that takes out endothermic negative energy and becomes cold when energized. Positive energy and negative energy is the energy that becomes zero when the equiponderance is added and mixed. Negative energy becomes a main stream from positive energy in space.

3.1 Use of Cold Current and Negative Area

Because the earth replaces a negative area in space by its mass, any negative area disappears. But because the methods are limited we assume the location to be negative. One method uses a permanent magnet, and another uses the Moebius ring.

3.1.1 Method with Permanent Magnet

The magnet is one of the three great inventions of our world. It was invented in ancient China, in the period "Spring and Autumn", for a car in instructions.   This vehicle had a compass whose needle was always pointing south. The mysterious nature of a magnet is that once it is magnetized by the pulse, its magnetism can be maintained for a long time.

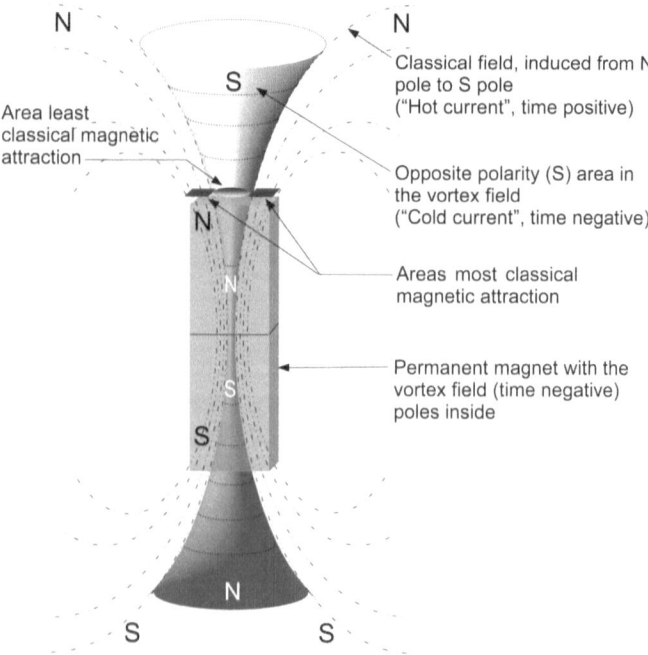

Figure 3.1.1 A New Concept of the Magnet[1]

This is a new concept of the magnet introduced by Tim Harwood and John Jankowski in Figure 3.1.1. The point is: until now it was explained that the bar magnet with N and S on both ends was induced from N pole and ended by entering S pole (Hot current, time positive). In addition, a magnetic field of the second whirlpool shows that there is S pole area (Cold current, time is negative) in the upper part of N pole. The location can be easily assumed to be a negative area when a magnet is used.

There are examples of using permanent magnets:
"Adams Motor" by Robert Adams, "Kepper Motor" by Norberto R. Keppe, "PODII" by Tim Harwood & John Jankowski, "SQM/VTA" by Floyd Sweet, "Mark III" by Lester Hendershot, "Magnet Current Apparatus" by Hans Coler, "M - L Converter" by Paul Baumann.

3.1.2 Method of Using the Moebius Ring

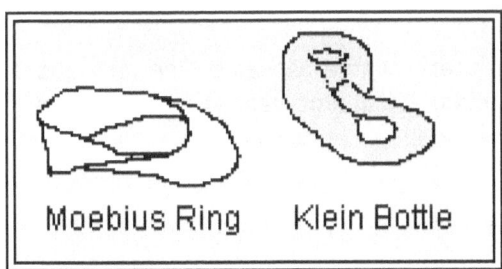

Moebius Ring   Klein Bottle

Figure 3.1.2-1 Moebius Ring and Klein Bottle[2]

Shinichi Seike introduced the method of using the Moebius Ring. By using his method, a strong electromagnet can be expected because resistance is near zero. The Moebius Ring (Figure 3.1.2-1) is made in a crowded three-pile layer and the insulator is placed between two metals. If the Moebius wounds are arranged to a torus (tube), they become the Klein coil. (Figure 3.1.2-2) The Moebius Ring changes electron to positron.

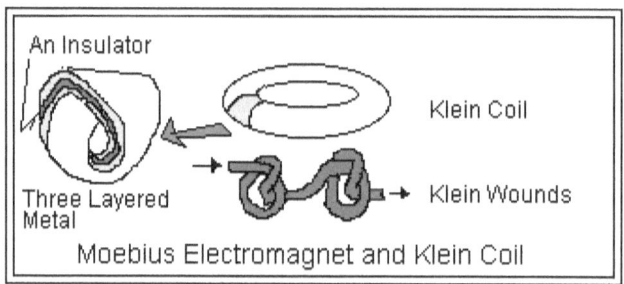

Figure 3.1.2-2 Moebius Electromagnet and Klein Coil[2]

The inside of the Klein coil assumes the positive area and the outside assumes a negative area. In Hans Coler's device of "Current Generator" (Figure 4.1):

Electromagnet M1 → Silver line → Copper plate P1 → Silver line → Electromagnet M2 → Silver line → Copper plate P2 → Silver line → Electromagnet M1

Because an electromagnetic charge passes on the backside of plate P2, if an electromagnetic charge passes the front side of plate P1, the loop is the Moebius ring rolling once.

3.1.3 Negative Area

• It is thought that the inside of plate P1/P2 is positive and the outside of plate P1/P2 opposed to plate F1 and F2 is a negative area (Figure 4.1).
• The first coil is a Moebius Ring in Nikola Tesla's "Tesla Coil" (Figure 4.7).
• It is a core technology of the free energy device to use the permanent magnet and of the Klein Coil to assume the place to be a negative area.

3.1.4 Summary

• In the flux of magnetic induction of the bar magnet that has N and S in two poles, it is an explanation of the flux of magnetic induction of the past magnet to end going out of N pole like the oval and entering in S pole (Hot and currently: A hot current and time is positive).

There is an area in N pole (A cold current and time is negative) in the upper part of S pole and S pole in the upper part of N pole placed between an oval flux of magnetic induction toward S pole from N pole.

• By putting the coil on a negative area of the magnet, the place becomes a negative area.

• By using the Moebius Ring the inside of the Klein coil is assumed to be a positive area and the outside is assumed to be a negative area.

• The Moebius Ring changes electron to positron.

3.1.5 Reference

1. http://www.angelfire.com./ak/energy21/adamsmotor.htm
   P22/33

2. http://www.geocities.co.jp/Technopolis/1123/seike.html

3.2 Increasing Power with Coil

1. The coil is an indispensable element for the free energy device. There is a solenoid coil of the electromagnet type to put the central core of iron and the midair type. If the central core of iron stays put, magnetism is strengthened, and when efficiency is valued because the loss of iron is generated, the midair type becomes primary.

2. The flux of magnetic induction flows to the center of the midair coil, which is shown in Figure 3.2.1. There is no magnetic field outside of the midair coil and the vector potential is generated. The first coil and the secondary coil of the transformer are exchanged by this vector potential.

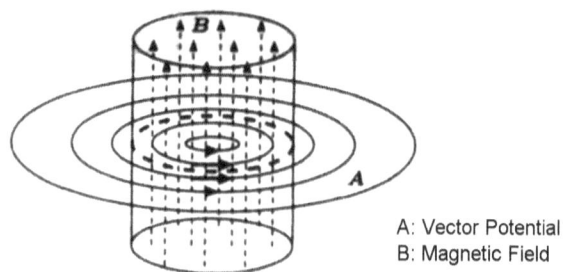

A: Vector Potential
B: Magnetic Field

Figure 3.2.1 Vector Potential[1]

3. The Bifilar Coil that Nikola Tesla invented (Patent number: US Pat 512,340) in Figure 3.2.2 is very famous. Bifilar Coil rolls two coils at the same time, and is a return of one coil on the output side to the other coil on the input side. The current will flow two times in the coil. The voltage between coils becomes 0.1 V per coil one roll by 100 V/1000 T= 0.1 V/ T at100 V1000 turn (T). It becomes 100 V/2 = 50 V for the Bifilar coil. The energy that can be accumulated in Bifilar coil usually becomes 50 V^2 /0.1 V^2= 2,500 / 0.01= 250,000 time compared with the normal coil.

The advantages of the Bifilar in this design are:
1. The Coil responds faster to the firing pulse
2. Increased Magnetic Field Strength produced
3. Increased Back EMF produced in field collapse
4. Increased Generator output

Because the magnetic field where another winding which is opposite and equal is made for the magnetic field with one winding, it becomes a magnetic field equal to zero. As a result, it becomes self-inductance equal to zero of the coils.

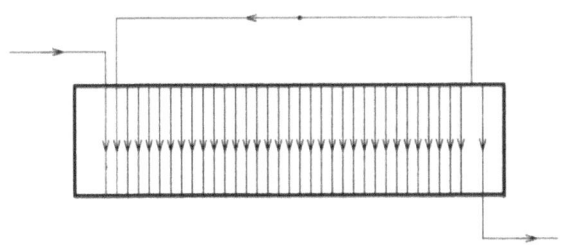

Figure 3.2.2 Bifilar Coil[2]

4. Bifilar coil P1, P2, F1 and F2 has been rolled in plate P1, P2, plate F1, and F2 in Hans Coler's "Current Generator" (Figure 4.1). The scalar potential is generated by bifilar coil.

5. There are power coil P1 and P2 used for "SQM/VTA" of Floyd Sweet's development (Figure 4.6.1) as a deformation of Bifilar coil. The current mutually flows in the opposite direction of the two coils, and the formed electric field has been canceled. The magnetic fields have canceled each other and become zero.

6. In Hans Coler's device of "Magnet Current Apparatus" (Figure 4.2.1) a single coil is increasing power according to the magnetism of the magnet as for the coil. The working is thought to be Bifilar coil equivalent.

7. Bifilar coil is rolled in a cylindrical magnet in "PODII" (Figure 4.4.1) as per Tim Harwood and John Jankowski. The increasing power is obtained with the Bifilar coil and the increasing magnetism using the magnet.

8. "Increasing power with coil" is a core technology in the free energy device.

3.2.1 Summary

•The flux of magnetic induction flows to the center of the midair solenoid coil. There is no magnetic field on the outside, and the vector potential is generated.

• The first coil and the secondary coil of the transformer are exchanged by the vector potential.

• Bifilar coil that Nikola Tesla invented rolls two coils at the same time, and one coil on output side is returned to the other coil on the input side.
   The Bifilar coil has 250,000 times cumulative capacity compared with the single coil.

•The scalar potential is generated by Bifilar coil.

•The coil P1/P2 of "SQM/VTA" device that Floyd Sweet invented is a deformation of Bifilar coil. The current flows to two coils in the opposite direction and the electric field has been canceled. The magnetic field where another winding which is opposite and equal is made for the magnetic field with one winding, it becomes a magnetic field equal to zero.

3.2.2 Reference

1. "Meaning and Concept of Electromagnetism," Yoichi Okabe, Kodansha Ltd., November 20, 2008 P92

2. http://overunity.infrance.com/Magnetricity_com

3.3 Use of the Complex field "Magnetic Field Cancellation with Coil"

1. The complex field of magnetic field cancellation with coil is famously known as the invention of Nikola Tesla (The patent number: US Pat 568,176).

From the equivalent circuit (Figure 3.3) according to the circuit explanation by William Alek of the United States, "Nikola Tesla was the first to develop by the phenomenon of complex fields back in the 1880s. He devised a series of machines patented in the 1890's that greatly amplify this phenomenon, which he later called RADIANT ENERGY.

As shown above, the pivoting magnetic domains created by Amperian Currents of the ferromagnetic material are ordered in the direction of field B_G by magnetizing coil G.

Magnetizing the high inductance coils M create an opposing field $\mu_0 H_M$ that acts upon the ordered domains of the material, thus canceling or partially canceling the *real* magnetic field created by the Amperian Currents.

An *imaginary* magnetic field jB_M emerges due to this cancellation and couples back into the magnetizing direct current as $i_M e^{j\theta}$, where $\theta > 0°$. Therefore, the magnetizing direct current becomes *complex* because the circulating motions of the electrons are rotating into the *imaginary* axis.

As shown above, before switch F is closed, the capacitors H are charged with a **complex direct current** $i_M e^{j\theta}$ produced by an opposing flux from coils M. The complex field energy is stored in capacitors H. At the moment of switch F closure t = 0sec, the **complex direct current** flows through coil K, rapidly discharging capacitor H. A very large *complex* electric potential $v_L e^{j\theta}$ is observed across the secondary coil L."

• It is a core technology in the free energy device that makes the complex current by canceling the magnetic field and obtains a very big amplification current by the complex potential. What is called Nikola Tesla Radient Energy is charge es^2 of Strong

Power, and Dark Energy.

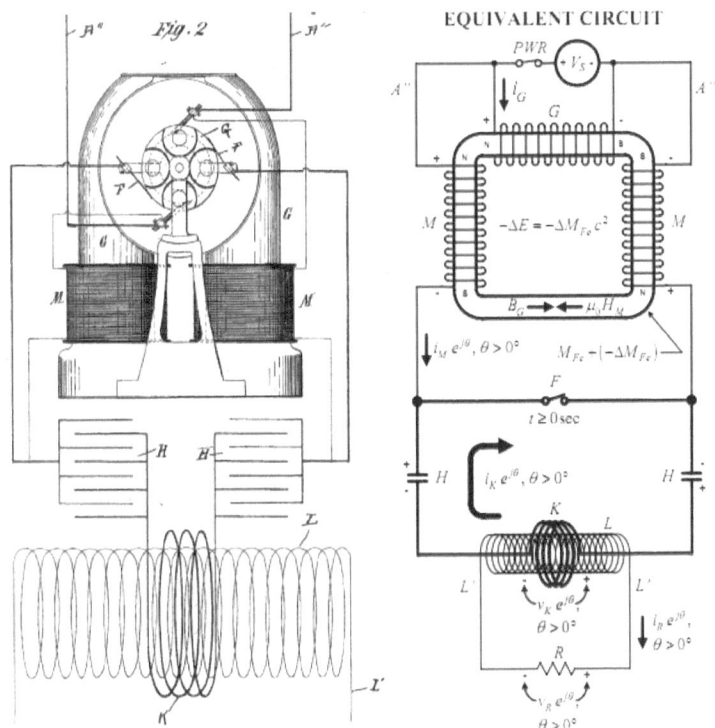

Figure 3.3 Tesla's Complex Field Generator[1]

2. The principle of the circuit of this patent is used in Hans Coler's "Magnet Current Apparatus" (Figure 4.2.1). Magnetic fields of coils two and five are canceled by the magnetic field of the magnet and are a complex current in the switch-off.

3. The magnetic field of coil F1(N-S) is canceled by the magnetic field of electromagnet M1(N-S) in Hans Coler's "Current Generator" (Figure 4.1) and becomes the complex current.

4. The magnetic field of BP2 coil (S-N) is canceled by the

magnetic field with magnet (N-S) in "Big Can" and becomes a complex current in Paul Baumann's "M-L converter" (Figure 4.5.1).

5. When the pulse is on condition the magnetic field of coil 2(N-S) and coil 3(N-S) is canceled by the magnetic field with cylindrical magnet (S-N), and it becomes a complex current in "PODII" (Figure 4.4.2) as per Tim Harwood and John Jankowski.

3.3.1 Summary

- By magnetic field cancellation with the coil that was invented by Nikola Tesla, a very big amplification complex current is obtained by the complex electric potential.
- Radiant Energy=Charge es^2 of Strong Power = Dark Energy

3.3.2 Reference

1. http://www.intalek.com/Index/Index.htm Newtonian Torsion Physics, William S. Alek INTALEK, INC., Page112

3.4 Use of Back Electromotive Force with Coil

1. Dr. Robert Adams developed Adams motor where the output produced is greater than the input, resulting for the first time in history the phenomenon of a motor becoming cold while operating. The motor is driven by using the pulse and generates back electromotive force in the coil of the device by the magnets that suppress the input electric power and rotate working.

2. Three coils are connected with the series, and the twice output/ input is obtained by using back electromotive force though it is used with "PODII" that Tim Harwood and John Jankowski developed (Figure 4.4.1).

3. William Alek is introducing the use of back electromotive force to the N. Zaev device[1].

4. Standing wave is formed with reflected wave and progressive wave in secondary coil of Nikola Tesla's "Tesla Coil" (Figure 4.7). It is possible to assume the reflected wave to be equivalent to the back electromotive force.

5. "Use of back electromotive force with coil" is a core technology in the free energy device.

3.4.1 Summary

> • The output that suppresses the input electric power can be obtained by using back electromotive force with coil.

3.4.2 Reference

1. http://www.intalek.com/Index/Index.htm Newtonian Torsion Physics, William S. Alek INTALEK, INC., Page107

3.5 Use of Pulse

1. The reverse electric power is caused by the direct current pulse of 4-5 kHz in the contact time of the length of 1/5–1/4 during the insulation time in Nikola Tesla's invention (The patent number: US Pat 577,670). This flows to the capacitor from over unity "Tesla Coil" conversion machine. The direct current pulse that unites with a negative impulse causes the back electromotive force to a high voltage. The midair type coil consequentially invents the effect of over unity of a high voltage (Figure 3.5).

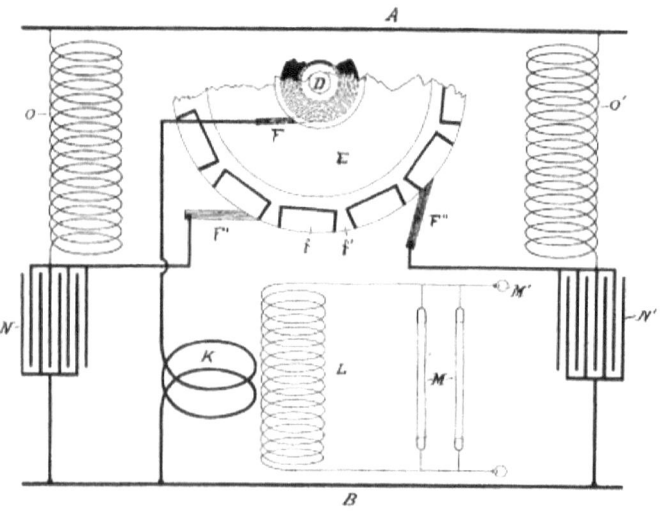

Figure 3.5 Use of Pulse[1]

2. The input pulse obtains twice the output for the input as a duty ratio of 1/4 in "PODII" that Tim Harwood and John Jankowski developed (Figure 4.4.1).

3. The reed switch "Turn On" works for a certain period by the disc rotation in Paul Baumann's "M-L converter" (Figure 4.5.2). It is estimated a method of the increasing power the same as use of pulse.

4. Spark discharge—here we see Nikola Tesla's "Tesla coil" (Figure 4.7) in the example of the representative that uses the spark discharge. It shortens capacitor for generating of wideband alternative pulse.

5. "Use of pulse" is a key technique in the free energy device.

3.5.1 Summary

> • We can ensure the increasing power energy continues operating by using the pulse.

3.5.2 Reference

1. "Night Before Energy Revolution," Kouya Mikami, Meisou publishing Inc. P178

3.6 Circuit for Making to Current of Electromagnetic Charge

1. The circuit for making to the current of an electromagnetic charge was invented by Hans Coler (Germany). Because he was a sailor, he was called Captain Coler. His results are open to the public as a part of British Intelligence Objectives Sub-Committee (B.I.O.S.) Trip No.2394 report, not by Germany but by Britain.

2. The current flows from S side to N side, and that is how the polarity of the mutual inductance of which it goes out to N side becomes negative to roll. There is a character that becomes negative in the mutual inductance insofar as resistance, capacity, and inductance never become negative polarities. There was no explanation in a past manual as to why a character of the mutual inductance becomes negative.

   It is Hans Coler who has given us the answer. It is explained by the "Current Generator" device in the above-mentioned report. Circuit (P1, P2) that rolls flat coil like litz wire (it can enlarge Q) to plate (it's made by copper) is composed as follows: P1 - magnet coil S pole - N pole – coil - magnet coil S pole - N pole - P2. The electromagnetic circuit has the mutual inductance of a negative way to roll and shows that it has the function to change an electromagnetic charge into the current (Figure 3.6).

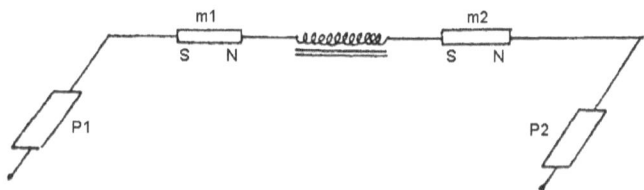

Figure 3.6 Circuit for Making to Current of Electromagnetic Charge[1]

3. This circuit "magnet coil S pole - N pole – coil - magnet coil S pole - N pole" is used for the free energy device "M-L converter" developed in Switzerland (Figure 4.5.1). The circuit is used for part "HP1, HS1, HP2, HS2" where the coil was rolled in the part of two U shaped magnets that change the collected static electricity into the current. A Big Can primary circuit BP1, part of magnet S pole – N pole and the coil, changes the electrical static charge into the current.

4. "Circuit for making current of electromagnetic charge" is a core technology in the free energy device.

3.6.1 Summary

> • Hans Coler developed the circuit "P1 - magnet coil S pole - N pole – coil - magnet coil S pole - N pole - P2".
> The electromagnetic circuit has the mutual inductance of a negative way to roll and shows that it has the function to change an electromagnetic charge into the current.

3.6.2 Reference

1. The Invention of Hans Coler, Relating to an alleged New Source of Power. British Intelligence Objectives Sub-Committee 28/07/2006

3.7 Output Circuit

1. There is a way that the current flows from N side to S side and the mutual inductance becomes positive to roll it in the coil. Flat spool circuit "F1, F2" is composed of a flat spool, and the name was taken from the appearance of the winder to roll the coil on the copper plate, in Hans Coler's report (Figure 3.7) as follows:

   F1 - magnet coil N pole - S pole - output coil - magnet coil S pole - N pole - F2

The electromagnetic circuit used for the output circuit is how the mutual inductance becomes positive and the mutual inductance becomes negative to roll. They show that it has the function to output the voltage.

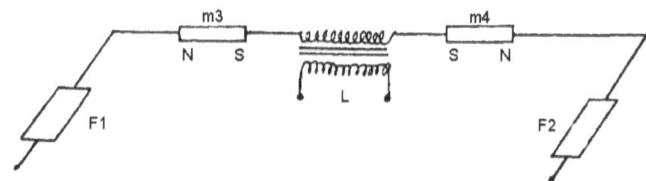

Figure 3.7 Output Circuit[1]

2. It is a composition in which all magnets and coils form the output circuit in the state of the switch-off in Hans Coler's devices "Magnet Current Apparatus" (Figure 4.2.2-1). The complex current is saved in the capacitor.

3. Hans Coler was making an important remark about the circuit design in the BIOS report when he said, "The resistance of the input is made ten times or more resist of the output."

4. As for "Output circuit", it is a key technique in the free energy device.

3.7.1 Summary

• Hans Coler's electromagnetic circuit is composed as follows: F1 - magnet coil N poles - S pole - output coil - magnet coil S pole - N pole - F2 used for the output circuit by how the mutual inductance becomes positive, and when the mutual inductance becomes negative to roll it is circuit that outputs the voltage.

• Hans Coler was recommending the resistance of the input to ten times or more resist of the output.

3.7.2 Reference

1. The Invention of Hans Coler, Relating to an alleged New Source of Power. British Intelligence Objectives Sub-Committee 28/07/2006

3.8 Resonant Circuit

1. The resonant circuit with the coil and the capacitor is usually used for the free energy device. Floyd Sweet invented the "SQM/VTA" device (Figure 4.6.1) in one of the strangest resonant circuits in a free energy device ever made. The display of the capacitor is not and is formed only with the coil in the schematic diagram. There is a reason for this: As an engineer in charge of the transformer Floyd Sweet was working on in the United States, General Electric Co. said that this is the company that originated from Thomas Edison's company. Therefore, he is quite a professional regarding the coil, having made the resonant circuit using the stray capacity between coils as a capacitor[1.]

2. There is an important frequency in the resonant circuit, found in a report in the thesis "Atom-Molecule in a Bose-Einstein Condensate" by Elizabeth A. Donley , Neil R. Claussen, and

others. There the double pulse of 20µs and 100µs is used.

The continuance oscillation by the standing wave of the atom and the molecule was observed in the experiment. Though pulse 20µs reaches frequency calculation value 50 kHz and it is 78 kHz and the electron is a fermion of spin 1/2, it actually becomes one wavelength with 2 pieces. It is shown that the actual measurement value is 157.8 kHz. Increasing power is thought to be activated by the electron in this frequency resonating and increasing the number of free electrons. The frequency in which the electron is activated seems to be common.

According to Mark III circuit chart by American inventor Lester Hendershot, it becomes f= 158 kHz(6.3µs) if it becomes L1= 120µH and C= 7800pF + 662.9pF= 8462.9pF and it calculates by this. The circuit of Floyd Sweet becomes P1 + P2= 156 kHz – 159 kHz[2].

3. The frequency calculation value of pulse 100µs is 9.77 kHz. The actual measurement value by the graph of pulse 100µs becomes 4.6 kHz in which frequency is an audio area and it is thought that it is the one to loosen molecular binding.
  As the molecule having mass, the transformation changes slowly and the frequency falls. The frequency in which the molecule is loosened seems to be common. According to Lester Hendershot's Mark III circuit chart, it becomes f= 4.6 kHz (217µs) if it becomes L3= 30µH and C= 40µF, and it is calculated by this. The circuit of Floyd Sweet becomes f= 4.6 kHz[3].

4. Hans Coler assumes the increasing power frequency of a free electron to be 180 kHz and the loosening frequency of molecular binding to be 10 kHz.

5. The resonant circuit (Figure 4.7) of Nikola Tesla's "Tesla Coil" is famous. The first coil and secondary coil are strongly united by the resonance.

As for the resonant circuit, it is a core technology in the free energy device.

3.8.1 Summary

• Among the resonance frequencies of the resonant circuit, there are two important kHzs. One is 158 kHz to activate the electron and to increase the number of free electrons. The other is 4.6 kHz to loosen molecular binding.

• In other words, it is important to use the double pulse of 6.3μs, and 217μs to continue oscillation by the standing wave of the atom, and molecule.

3.8.2 Reference

1. Essay "Window" the 17th collection "Resonance Frequency", Makoto D. Yamane, Meisou Publishing Inc. Dec.20th, 2010 P184

2. http://www.hyiq.org/Library/Floyd_Sweet.htm   April 28, 2005 material
http://www.library.uu.nl/digiarchief/dip/diss/2003-1104-110733/c6.pdf

3. "The Hendershot Mystery," Secret of Perpetual Power, Victoria Australia, 1998 Printed, published and distributed by NUTECH 2000

3.9 Oscillation Circuit

1. The function of small oscillator (OSC) has been added to the magnet, which is described in the thesis "Nothing is Something" in SQM/VTA by Floyd Sweet (Figure 4.6.1). When magnetizing, this memorizes an excessive signal of 60 Hz in the magnet.

  This signal works respectively as a pulse, giving the increasing power energy for operation to continue to output coil P1, P2, feedback coil FB1, FB2, and works so that the frequency of the output coil may be steady[1].

Exchange of 400 Hz with Floyd Sweet contribution SQM Mark II is added as an input signal. "It is described that the quality of the input oscillator is important." Making this part of the black box is measured by spending the greater part of time in the development of the method for memorizing this oscillator (60 Hz) signal in the magnet after the success of the experiment.

2. The oscillatory frequency of 15cm and 90 kHz × 2= 180 kHz is obtained by using the resonance frequency of iron by the central core of iron with electromagnet M1 and M2 in Hans Coler's "Current Generator" device (Figure 4.1). Hans Coler describes having stored ten oscillation circuits in this circuit[2].

It is thought to have the following content:
    1. 180 kHz by central core of iron with electromagnet M1
    2. 180 kHz by central core of iron with electromagnet M2
    3. Frequency 10 kHz of coil with electromagnet M1
    4. Frequency 10 kHz of coil with electromagnet M2
    5. The frequency of the magnetic field in the central core of iron loop with magnet M1 is 180 kHz:

    Central core of iron M1 → Silver line → Plate P1 → Central core of iron M2 → Silver line → Plate P2 → Silver line → Central core of iron M1

    6. The frequency of the electric field in the coil loop of magnet M1 is 10 kHz of the vector potential:

    Coil of M1 → F1 coil → Coil of M2 → F2 coil → Coil of M1

    7. Resonance frequency 180 kHz of coil P1
    8. Resonance frequency 180 kHz of coil P2
    9. Resonance frequency 10 kHz of coil F1
    10. Resonance frequency 10 kHz of coil F2

3. Hans Coler assumes the frequency by increasing the number of free electrons to 180 kHz and the frequency by loosening molecular binding to 10 kHz. It is thought that 156 kHz – 159 kHz and 4.6 kHz is better from other cases.

4. The oscillatory frequency   "180 kHz" (Is it 158 kHz accurately?) is obtained by adjusting the central core length with iron 10 cm = 60 kHz and U shape type magnet + electromagnet CH1 + electromagnet CH2 in Lester Hendershot's MarkIII (Figure 4.3.1).

5. The time constant is set to vacuum triode tube with coil L1, variable capacitor VC4, and capacitor C5 after it becomes a reed switch on and the oscillator is formed in Paul Baumann's "M – L converter" (Figure 4.5.1).

6. The AC power frequency is assumed to be a frequency of the oscillator in Nikola Tesla's "Tesla Coil" (Figure 4.7).

7. "Oscillation circuit" is a key technique in the free energy device.

3.9.1 Summary

• The oscillator is built into the device, if not built into it, the AC power frequency is assumed to be a frequency of the oscillator.

3.9.2 Reference

1. "Nothing is Something," Floyd A. "Sparky" Sweet, June 24th, 1988

2. The Invention of Hans Coler, Relating to an alleged New Source of Power. British Intelligence Objectives Sub-Committee 28/07/2006

Chapter 4

Circuit Examples with Free Energy Devices

There are free energy devices that have large amounts of excess free energy. Despite the lack of sufficient materials, I will try to analyze these circuit operations.

The following examples are typical free energy devices.

---

1. "Current Generator" by Hans Coler

2. "Magnet Current Apparatus" by Hans Coler

3. "Mark III Solid State Generator" by Lester Hendershot

4. "Power On Demand II (Pod II)" by Tim Harwood & John Jankowski

5. "Methernitha-Linden (M-L) Converter" by Paul Baumann

6. "Space Quanta Modulator (SQM) / Vacuum Triode Amplifier (VTA) " by Floyd Sweet

7. "Tesla Coil" by Nikola Tesla

Warning

These circuits are prepared for explanation and not guaranteed for operation. As for frequency, I have entered a numerical value as the material permits. When no materials are available, I put a preferable frequency.

When making trial pieces, please guarantee your own safety, knowing that you are solely responsible.

4.1. "Current Generator" by Hans Coler

4.1.1 Analysis Result

1. Switches are on in the order of SW1, SW2, and SW3 (Figure 4.1). The current in electromagnetic coil M1 and M2 flows, and they work as a solenoid after switch one is on. As for the central core iron of electromagnet M1 and M2, resonate to180 kHz and for coil M1 and M2, resonate to 10 kHz by L and stray capacity C of the coil.

2. Flux of magnetic induction "180 kHz" from the central iron core of Electromagnetic M1 turns:

M1 N pole →Silver line → Plate P1 → Silver line → M2 S pole → M2 N pole → Silver line → Plate P2 → Silver line → M1 S pole → M1 N pole.

Vector potential "10 kHz" is generated outside of coil M1 and M2 and the difference of an electromagnetic potential happens between plate P1 and P2.

3. Coil P1 and P2 rolled in plate P1 and P2 change the charge accumulated between plate P1//plate F1 and plate P2//plate F2 by the vector potential generated for electromagnet M1 and M2 into the current. The vector potential loop is like this:

Electromagnet M1(S→N) → Coil P1(S→N) → Electromagnet M2(S→N) → Coil P2(S→N).

This circuit is Hans Coler's invention where an electromagnetic charge creates the current. The P1, P2 coils have been rolled anti-clockwise in view of N side up and their mutual inductance becomes negative.

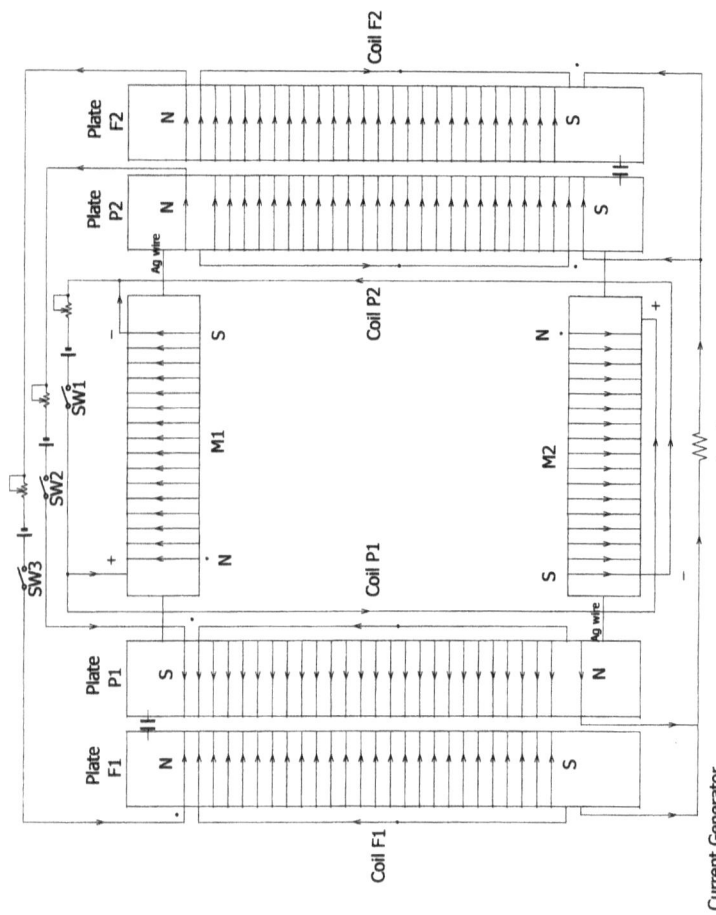

Figure 4.1 Current Generator

4. Because the flux of magnetic induction flows on the front side of plate P1 and backside of plate P2, both plates become the Moebius ring of rolling once. The inside of P1 and plate P2 are placed to positive areas and both outside of plate P1 opposed to plate F1 and outside of plate P2 opposed to plate F2 are placed into a negative area.

5. Coil P1 and P2 resonate from inductance L and stray capacity C of the coil to 180 kHz and increase the number of free electrons. Moreover, coil P1 and P2 cause the scalar potential for the bifilar wound and the current is increasing power.

6. Plate F1//plate P1 and plate F2//plate P2 form the capacitor. The transformer unites with coil F1, P1, and coil F2, P2. Coil F1 and F2 do work that resonates from inductance L and stray capacity C of the coil to10 kHz, loosening molecular bindings, and increase the number of atoms.

Moreover, coil F1 and F2 interfere and increase the number of free electrons with coil P1 and P2. The magnetic field is canceled by the flux of magnetic induction of electromagnet M1 and M2 and coil F1 is made a complex current. In coil F2, the current increases in power according to the flux of magnetic induction of electromagnet M1 and M2. Coil F1 and F2 cause the scalar potential for the bifilar wound and cause the current to increase power. Coil F1 and F2 make the following output circuit:

Plate F1 - F1 coil N pole - S pole - output coil - F2 coil S pole - N pole - Plate F2.

   Coil F1 mutual inductance becomes positive and coil F2 mutual inductance becomes negative to roll they show that it has the function to output the voltage.

7. The increasing power of complex current flows and the load is driven from the output of coil F1 and P1 through the output of coil P2 and F2.

4.1.2 Discussion of Maxwell's Electromagnetic Potential Equation

1. $-E = \nabla \cdot \phi + \partial A / \partial t$

The electric field becomes negative (negative area) by the time change of the emanation of the scalar potential "coil P1, P2, F1, and F2 bifilar wound" and the vector potential "Go out of the outside of electromagnet M1 and the M2 coil."

2. $B = \nabla \times A$

The magnetic induction rises by rotating the vector potential (electromagnet M1 coil→ Coil P1 → Electromagnet M2 coil → Coil P2 → Electromagnet M1 coil).

3. $\partial \rho / \partial t = \nabla \cdot j$

The current diverges by the time change of the charge "capacitor of plate P1//plate F1 and plate P2//plate F2" occurs. Coil P1 and coil P2 make the charge to give current.

4.1.3 Reference

1. The Invention of Hans Coler, Relating to an alleged New Source of Power. British Intelligence Objectives Sub-Committee 28/07/2006

4.2 "Magnet Current Apparatus" by Hans Coler

4.2.1 Analysis Result

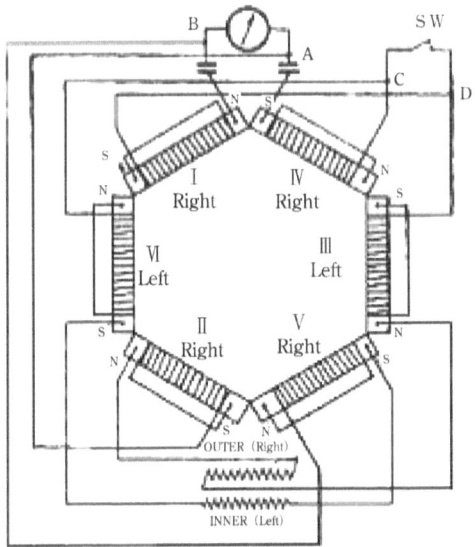

Figure 4.2.1 "Magnet Current Apparatus[1]"

1. The interval of the magnet is adjusted to about 7 mm in condition of switching off. It makes between magnets enter a negative area, and the flux of magnetic induction is adjusted to rotate through each magnet in Figure 4.2.1.

2. K coil "Outside" indicates "Right" to the rotation convergence, and K coil "Inside" indicates "Left" to the divergence in Figure 4.2.2-1. The current flows to:

K coil "Inside" edge N from inside center S → Magnet 6 (S→N) → Coil 6 (S→N), "Increasing power" → Coil 4 (N→S), "Increasing power" → Magnet 4 (N→S) →Capacitor CA (+).

3. The current flows to capacitor:

CB (-) → Magnet 1 (N→S) → Coil 1 (N→S), "Increasing power" → Coil 3 (S→N), "Increasing power" → Magnet 3 (S→N) → K coil "Outside" edge N → Center S.

4. The current flows to capacitor:
CA (-) → Coil 2(N→S), "Magnetic field cancellation" → Magnet 2 (S→N) → K coil (outside) edge N → Center S.

Because the magnetic field is canceled, the current becomes a complex current.

5. The current flows to K coil (inside):
Center S → Edge N → Magnet 5(S→N) → Coil 5(N→S), "Magnetic field cancellation" → Capacitor CB (+).

Because the magnetic field is canceled, the current becomes a complex current.

6. The central portion of K coil comes to zero point. K coil outputs the signal with the phase lag of -45°, +45° and becomes a Quadrature hybrid circuit[2].

7. The adjustment of putting the magnet position and K coil in and out is done so that the voltmeter voltage may be maximized (Figure 4.2.2-1).

8. The complex current is accumulated in each capacitor by turning on switch (SW) in Figure 4.2.2-2 mixes, and the current flows to capacitor:

CA (+) → Magnet 4(S→N) → Coil 4(N→S) "Magnetic field cancellation" → SW → Coil 1(N→S) "Magnetic field cancellation" → Magnet 1(S→N) → Capacitor CB (-).

The current of phase 45°and -45°mixes.

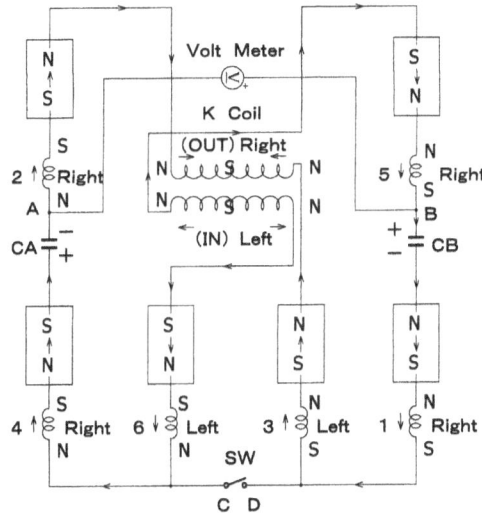

Figure 4.2.2-1 Magnet Current Apparatus

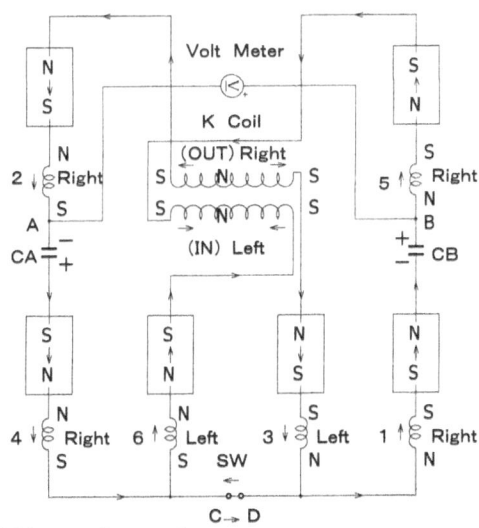

Figure 4.2.2-2 Magnet Current Apparatus

9. The following explains what happens in the capacitor:

CB (+) → Coil 5(N→S), "Increasing power" → Magnet 5(N→S) → K coil (inside) edge S → Coil (inside) center N → K coil (outside) center N→ Edge S → Magnet 3(N→S) → Coil 3(S→N), "Magnetic field cancellation" → SW →Coil 6(S→N), "Magnetic field cancellation" → Magnet 6(N→S) → K coil (inside) edge S → K coil (inside) center N → K coil (outside) center N → Edge S → Magnet 2(N→S) → Coil 2(N→S), "Increasing power" → Capacitor CA (-).

A very big complex current flows because of the magnetic field cancellation with Chapter 3 element technological coil the voltmeter voltage becomes the maximum.

4.2.2 Discussion of Maxwell's Electromagnetic Potential Equation

1. $-E = \nabla \cdot \phi + \partial A/\partial t$
The electric field becomes negative (negative area) by the divergence of the Scalar potential (rolling 1, 2, 3, 4, 5, and 6 coils are equivalent to the bifilar coil) and by the time change of the vector potential (K coil).

2. $B = \nabla \times A$
The magnetic induction rises by rotating the vector potential (coil 5, K coil "inside", K coil "Outside", coil 3, coil 6, K coil "Inside," K coil "Outside," and coil 2) at a switching on.

3. $\partial p/\partial t = \nabla \cdot j$
The current divergence emanates by the charge (capacitor CA and CB). The switch "On" makes the charge to current with coil four and coil one.

4.2.3 Reference

1. The Invention of Hans Coler, Relating to an alleged New Source of Power. British Intelligence Objectives Sub-Committee 28/07/2006

2. Revision new publication text "Utilization of Encyclopedia Toroid Core," Yamamura Hideho, CQ Shuppan Co., Ltd, March 1, 2007 P386

4.2.4 Clipping Data

•The center of K coil of "Magnet Current Apparatus" that Hans Coler developed is zero point. It has been understood so far from this:
Zero Point Energy (ZPE) = →Strong Power Charge es^2← = Dark Energy.

4.3 "Mark III Solid State Generator" by Lester Hendershot

4.3.1 Analysis Result

1. The central core of iron with electromagnet CH1, CH2, and iron bar are arranged in a negative area that exists in the tip of N pole and S pole of the U shape magnet in Figure 4.3.1.

2. U shape magnet, iron bar and two electromagnets CH1 and CH2 called a bell circuit create a magnetic bunch style caused by the magnet. An electromagnetic charge passing electromagnetic coil CH1 and CH2 is changed to a current with the circuit that was invented by Hans Coler.

3. Electromagnetic coil CH1 and CH2 resonate by "158 kHz" with the current's flow, increasing the number of free electrons. The magnet sets the length as resonating by "158 kHz."

4. The central core of iron with electromagnet CH1 is initially magnetized by magnet N pole. The magnet side becomes N pole and the other side becomes S pole. The current flows by the magnetic field from the N pole to S pole as follows:

Electromagnetic coil CH1(1) S → (2) N, and electromagnetic coil CH2(3) S → (4) N

5. Electromagnetic coil CH1(2) on the part of N pole of the magnet becomes N pole, and electromagnetic coil CH2(4) on the part of S pole of the magnet becomes N pole. The iron bar from N pole of electromagnetic coil CH2(4) is attracted to S pole of the electromagnet so that the flux of magnetic induction may head for magnet S pole. The iron bar is located in between on the part of magnet N pole and N pole of electromagnet CH1(2).

6. The back electromotive force is generated in electromagnetic coil CH1 and CH2 as in Figure 4.3.2.

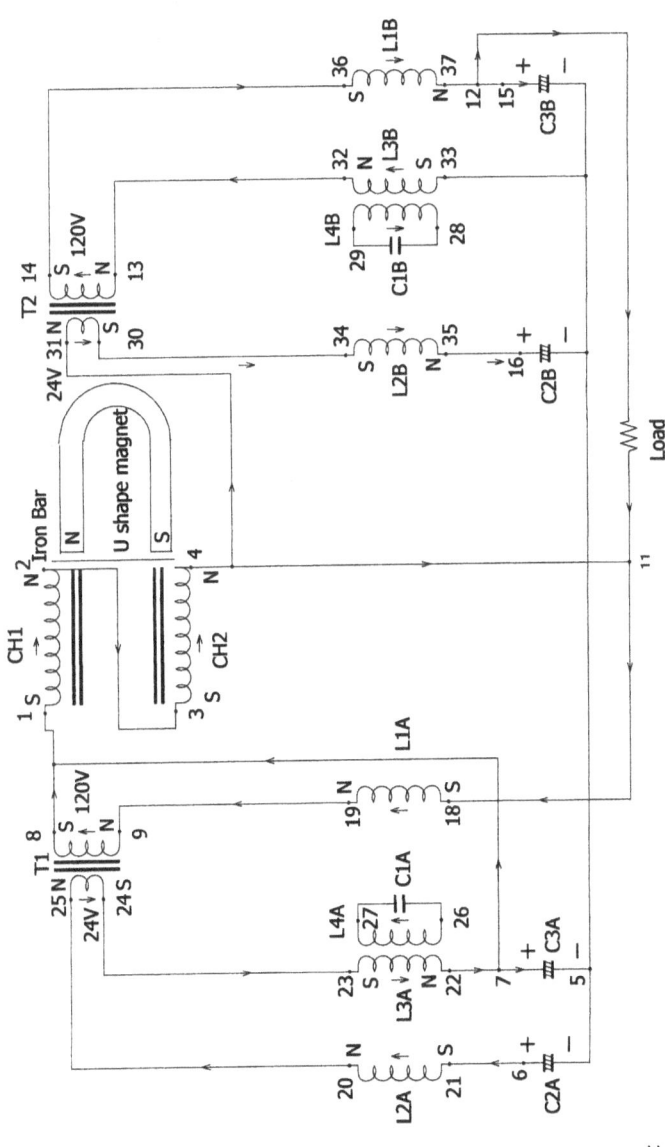

Start

Figure 4.3.1 MarkⅢ Solid State Generator

The current flows with CH2(4) S → (3) N, and CH1(2) S →(1)N. In order for the flux of magnetic induction to head from magnet N pole for electromagnet CH1(2) S pole through the iron bar, the iron bar is attracted to magnet N pole.

7. The iron bar starts the mechanical vibration by low frequency with a repeat of items five and six. Because the current flows to electromagnetic coil CH1 and CH2 flows to alternate CH2 and CH1, the coil of electromagnet CH1 and CH2 are resonated by "158 kHz." The current flows to electromagnetic coil CH1(1) S → (2) N, because magnet S pole → N pole and the polarity become opposite, the magnetic field is canceled and becomes a complex current.

On the other hand, electromagnetic coil CH2 becomes (3) S → (4) N. It is increasing power because the flux of magnetic induction becomes the same direction as magnet S → N.

8. Resonance frequency "158 kHz" of electromagnetic coil CH1 and CH2 is modulated by the low frequency of iron bar with item seven and the complex current is increasing power.

9. The current flows from item four with electromagnetic coil:

Electromagnetic coil CH1(1) S→(2) N→ Electromagnetic coil CH2(3) S → (4) N → (11) → L1A(18) S → (19)N → T1(9) N → (8) S → Electromagnetic coil CH1(1) S in Figure 4.3.l.

10. The back electromotive force flows from item six with electromagnetic coil:

Electromagnetic coil CH2(4) S→(3) N → Electromagnetic coil CH1(2) S → (1) N → T1(8) N → T1(9) S → L1A(19) S → (18) N → (11) → Electromagnetic coil CH2(4) S in Figure 4.3.2.

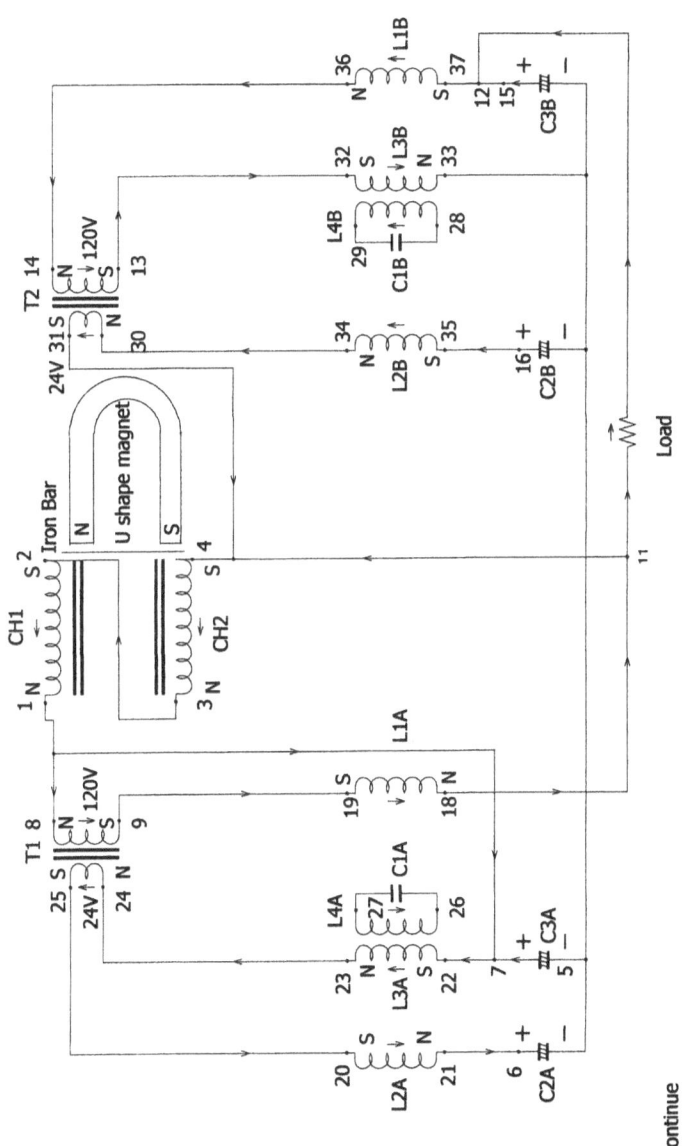

Continue

Figure 4.3.2 Mark III Solid State Generator

11. The voltage of transformer one rises by the repetition in Figure 4.3.1 and T1(8) becomes 0 → 120 V.

12. It interferes with L2A "4.6 kHz" frequency in which molecular binding is loosened and the number of atoms is increased, and with C1A "158 kHz" frequency in which the number of free electrons is increased from the atom, the charge is saved. This charge is changed into the current by electromagnetic coil CH1 and CH2.

13. The current flows to the second side of T2 as in Figure 4.3.1:

CH2(4) N → T2(31) N →(30) S → L2B(34) S →(35) N.

C2B is charged to 24 V. It interferes CH2 158 kHz and L2B 4.6 kHz, the number of free electrons is increased, and the charge is accumulated.

14. The current flows as in Figure 4.3.1:

L3B(33) S →(32) N → T2(13) N →(14) S → L1B(36) S →(37) N.

It is interfered by L4B 158 kHz and L1B 4.6 kHz, which cause the number of free electrons to be increased, and the charge is accumulated in C3B. T2(14) rises to 0 → 120 V.

15. The current flows to the second side of T1 as in Figure 4.3.2:

C3A(7) (+) → L3A(22) S → (23) N → T1(24) N → (25) S → L2A(20) S → (21)N.

C2A is charged. The number of free electrons is increased to interfere with L4A 158 kHz and L2A 4.6 kHz and the charge is accumulated in C2A.

16. The current flows as in Figure 4.3.2:

C2B(16) (+) → L2B(35) S → (34) N → T2(30) N → T2(31) S →

CH2(4) S → (3) N → CH1(2) S → (1) N.

The charge is made into a current by electromagnetic coil CH1 and CH2.

17. The current flows to the primary side of T2 as in Figure 4.3.2:

C3B (+) → L1B(37) S → (36)N → T2(14) N → (13) S → L3B(32) S → (33) N.

T2(13) rises to 0 → 120 V. It interferes with L4B 158 kHz and L1B 4.6 kHz, and the number of free electron is increased, and it steps up the voltage by T2.

18. It becomes 120V as for T1(8), 120V as for C3A(7), 24V as for CH2(4), and 120V as for T2(14) as in Figure 4.3.1. 24V and 120V are added to both ends of the load and the voltage of the substance 96V is impressed on the load.

19. It becomes 120V as for T1(9), 24V as for CH2(4), and120V as for T2(13) as in Figure 4.3.2. The voltage of -24V and -120V join the load. The voltage of the substance-96V is impressed on the load.

20. The alternating voltage of 96 V is impressed from item 18 and 19 on the load because it becomes 1/2 cycles.

4.3.2 Discussion of Maxwell's Electromagnetic Potential Equation

1. $-E = \nabla \cdot \phi + \partial A / \partial t$
The electric field becomes negative by the divergence of the scalar potential (coil L3A, L4A, L3B, and L4B) and by the change of time of the vector potential (coil L1A, L2A, L1B, and L2B).

2. $B = \nabla \times A$
The magnetic induction rises by rotating the vector potential (coil CH1, CH2, L1A, and 1st side of T1).

3. $\partial\rho/\partial t = \nabla \cdot j$

The current is diverged by a change of the time of the charge (the capacitor 3A and 3B). (The charge changes to current by coil CH1 and CH2.)

4.3.3 Warning (Feature of Complex Current)

Protection against the overload is necessary for this circuit.

There is a character to maintain the match to resistance about the complex current when it characterizes and a large current is thrown once.

This is because the device might be damaged with the current capacity that can be output even by the overload.

4.3.4 Reference

1. "The Hendershot Mystery" Secret of Perpetual Power, Victoria Australia, 1998 Printed, published and distributed by NUTECH 2000

4.4 "Power On Demand II (Pod II)" by Tim Harwood & John Jankowski

4.4.1 Analysis Result

1. A positive pulse inputs as in Figure 4.4.1 and Figure 4.4.2. 3A on the input side diode, and the current flows like coil 3(N→S), coil 2(N→S), and coil 1(N→S), etc. Because the magnetic field is canceled as a result of coil 3 becoming a polarity S, N, and a reverse-polarity of a cylindrical magnet, and coil 2 becomes polarity S, N, and a reverse-polarity of a cylindrical magnet, the magnetic field is cancelled and becomes a complex current (cold current).

Coil 1(N→S) is increasing power according to the Bifilar Coil because it is composed so that the head of the iron nail may enter a negative area of round type magnet S, becoming magnetization S, N, and the same direction with an iron nail magnet.

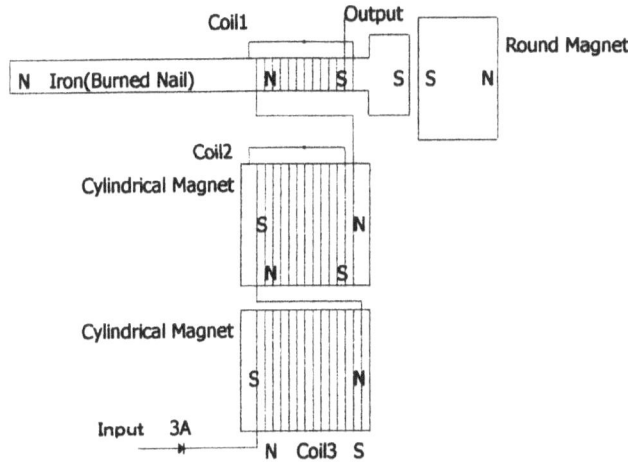

Figure 4.4.1 Pod II Circuit

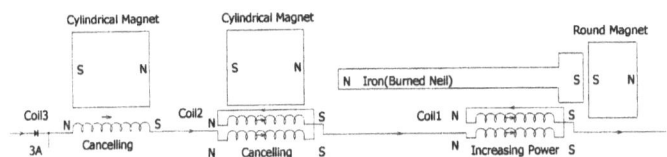

Figure 4.4.2 Pulse On

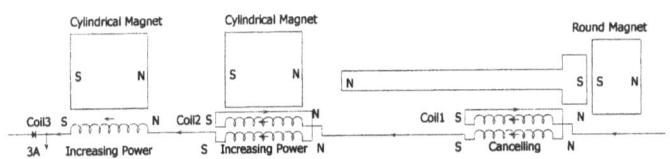

Figure 4.4.3 Pulse Off

2. When the pulse on ends (The pulse is off), the reverse current flows to coil 3 like coil 1(N→S), coil 2(N→S) and coil 3(N→S) because of the back electromotive force is caused in each coil in Figure 4.4.3 reverse direction to the item one.

The magnetic field is canceled because of the increasing power and coil 1 becomes magnetization N reserve of iron nail, which is the central core of iron S, and moves in an opposite direction. This is because the polarity of a cylindrical magnet becomes the same polarity as S and N, since the polarity of a cylindrical magnet coil 3 becomes the same polarity while increasing power. At the same time, coil 2 and the complex current (cold current) flows.

4.4.2 Reference

1. http://www.angelfire.com./ak/energy21/adamsmotor.htm   P19 /33

4.5 "Methernitha - Linden (M - L converter)" by Paul Baumann

4.5.1 Analysis Result

1. Two disc rotation beginning (manual operation)
1.1 The current from which static electricity is made up of a complex current and static electricity flows as in Figure 4.5.1:

ML2 (+) → C2(+) → Coil HS2(S→N) "Magnetic field cancellation" → U shape magnet (S→N) → U shape magnet (S→N) → Coil HP2(N→S) "Increasing power" → C3(-)→ BP2(S→N) "Magnetic field cancellation" → SC2 → C2(-)

1.2 The current from which static electricity is made up of a complex current and static electricity flows:

C1 (+)→ SC1 → BP1(S→N) "Increasing power" → Coil 3(+)→ HS1(S→N) "Magnetic field cancellation" → U shape magnet (S→N) → U shape magnet (S→N) → Coil HP1(N→S) "Increasing power" → C1 (-)

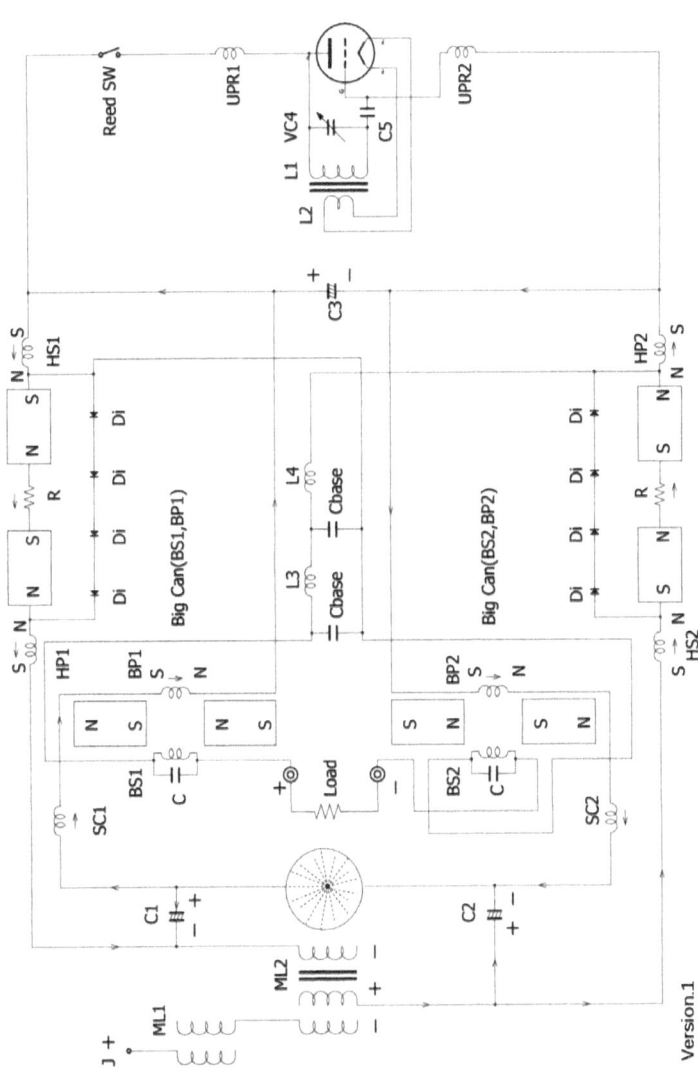

Version.1

Figure 4.5.1 Two Disc Rotation of Beginnig Of M-L Converter

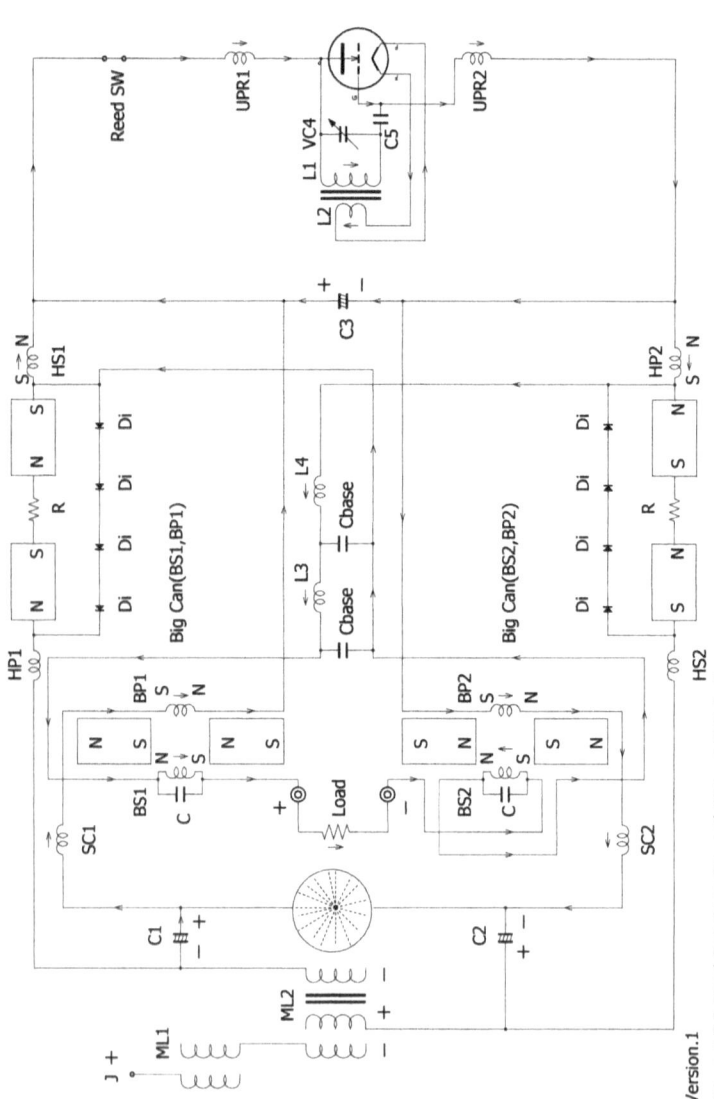

Version.1

Figure 4.5.2 Reed Switching On, and Vacuum Triode On

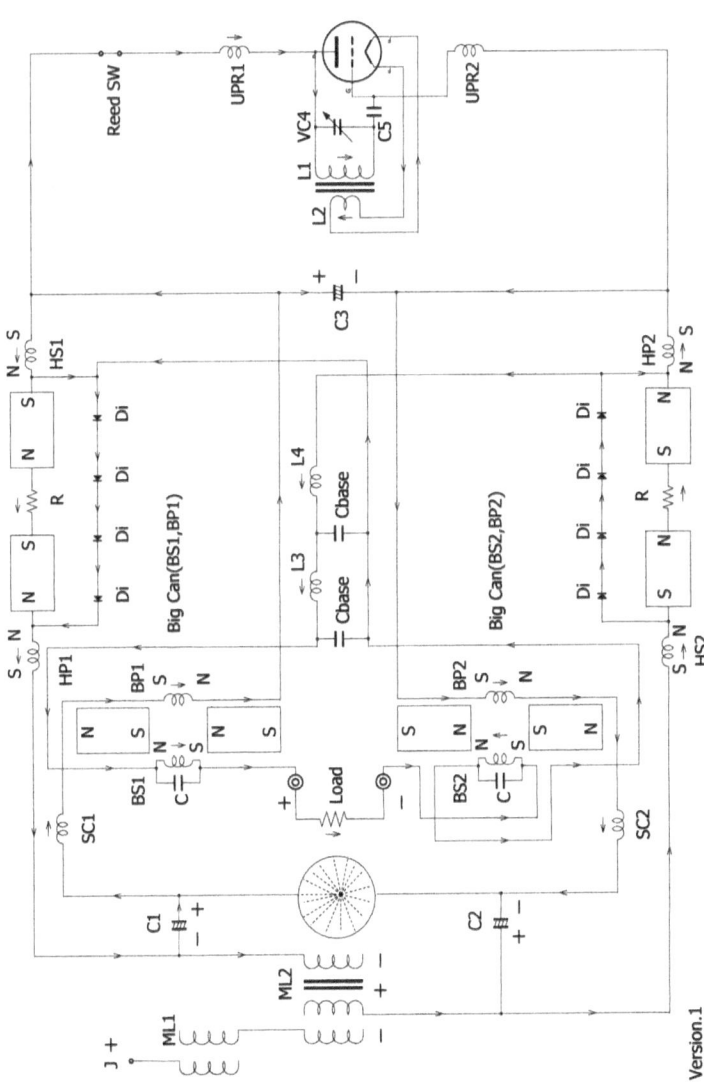

Version.1

Figure 4.5.3 Reed Switching On, and Vacuum Triode Off

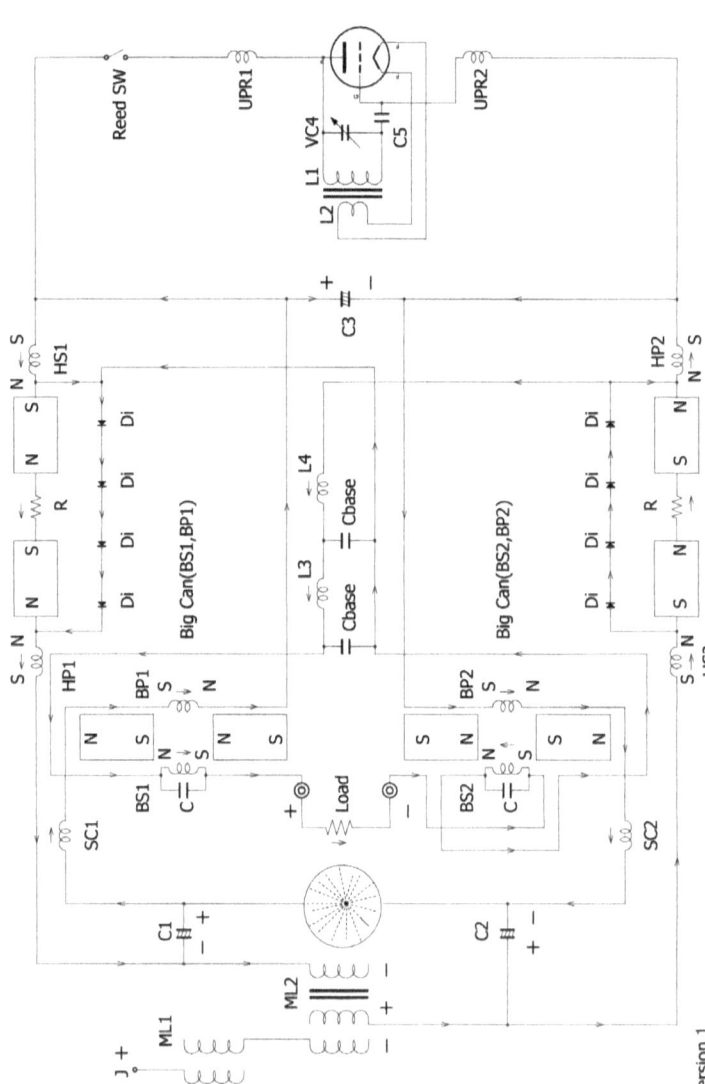

Version.1

Figure 4.5.4 Reed Switching off, and Vacuum Triode Off

2. These three coils are anti-clockwise rolled with the circuit in view of N side. They have negative mutual inductance and they change electromagnetic charge to a current.

2.1 Coil HS2(S→N) → U shape magnet (S→N) → U shape magnet (S→N) → Coil HP2(N→S)

2.2 Coil HS1(S→N) → U shape magnet (S→N) → U shape magnet (S→N) → Coil HP1(N→S)

2.3 BP1(S→N)

3. By rotating the disc of the device in Figure 4.5.2 the reed switch is on. The current flows to L1 and L2 and filament F heats the vacuum triode tube. By the oscillatory frequency of VC4, C5 and L1 the current "UPR1 → vacuum triode tube → UPR2" flows from vacuum triode tube plate P to grid G.

3.1 Because the vacuum triode tube operates like having put the low resistance for the complex current flows like:

C1 (+) → SC1 → BP1(S→N) "Increasing power" → C3(+)*→ Reed switch → UPR1 → Vacuum triode tube → UPR2 → C3(-)→ BP2(S→N) "Magnetic field cancellation" → SC2 → C2 (-)

3.2 The complex current flows like:

HP2(N→S) → L4, Cbase → L3, Cbase → BS1(N→S) "Magnetic field cancellation" → Output (+)→ Load → Output (-)→ BS2(S→N) → Cbase → Cbase → HS1(S→N) →*.

Coil BP1, BS1, BP2 and BS2 are in a negative area between upper and lower magnet.

4. Vacuum triode tube off.
Vacuum triode tube does not operate having put the low

resistance but the complex current continuously flows to coil BP1, BS1, BP2 and BS2 in Figure 4.5.3. The complex current flows from coil HP2, HS1 like:

ML2 (+)→ C2(+)→ Coil HS2(S→N) "Increasing power" → Four strings diode → L4, Cbase → L3, Cbase → BS1(N→S) "Magnetic field cancellation" → Output (+)→ Load→Output (-)→ BS2(S→N) → Cbase → Cbase → Four strings diode → HP1(N→S) → C1 (-) → ML2 (-).

If large complex currents flow, it is considered to be the feature of the complex current that flows continuously according to the load.

5. Reed switch-off.
The complex current continuously flows to BP1, BS1, BP2 and BS2 because the state of item four (Figure 4.5.3) continues even if the reed switch turns off as in Figure 4.5.4. Static electricity is charged to C2 (+), C1 (+).

6. The vacuum triode tube is an improvement over the one created by the increasing current power from the spark discharge. The increasing current power by the pulse is done for the period with the disc one rotation.

4.5.2 Discussion of Maxwell's Electromagnetic Potential Equation

1. $-E = \nabla \cdot \phi + \partial A/\partial t$
The electric field becomes negative by the divergence of the scalar potential (coil BP1, BS1, BP2 and BS2) and by the time change of the vector potential (coil HS1, HP1, HS2 and HP2).

2. $B = \nabla \times A$
The magnetic induction rises by rotating the vector potential (coil HS1, UPR1, UPR2, HP2, L4, L3, BS1 and BS2) when the vacuum triode tube is on.

3. $\partial\rho/\partial t = \nabla \bullet j$

The current divergence by the time change of the charge (capacitor C1, C2, and C3). (Electrical static charge changes to current by coil HS1, HP1, HS2, HP2 and BP1.)

4.5.3 Reference

1. Electric circuit as understood from original photos of the Methernitha Testatika by Paul Potter. Retrieved from http://www.liux-host.org/energy/fig.gif There are figures up to 1~7 & circuit.

4.6 "Space Quanta Modulator (SQM)/Vacuum Triode Amplifier (VTA)" by Floyd Sweet

4.6.1 Analysis Result

1. The current with the battery flows to the T1/ T2 coil becomes a midair solenoid when the power switch is adjusted to the on side in the schematic diagram of Figure 4.6.1. Flux of magnetic induction "35.1 kHz" is formed to the center, and the vector potential is generated outside of the coil.

The center flux of magnetic induction with T1 / T2 coil works in both N sides of power P1 / P2 coil. The magnetic density B from upper to lower magnet is squeezed by the center flux of T1 / T2 coil similar to a whirlpool and rises.

This looks like a towel containing water as one's right and left hand wring it dry. The alternative output is generated by the whirlpool of the flux of magnetic induction and the zoom of the magnetic induction in power coil P1 / P2 "158 kHz".

2. On the other hand, the vector potential generated outside of the T1 / T2 coil generates the charge by the barium ferrite magnet "Both high dielectric constant materials of barium and the ferrite." This charge acts on feedback coil FB1 and FB2 "316 kHz" that exists in the resonant case.

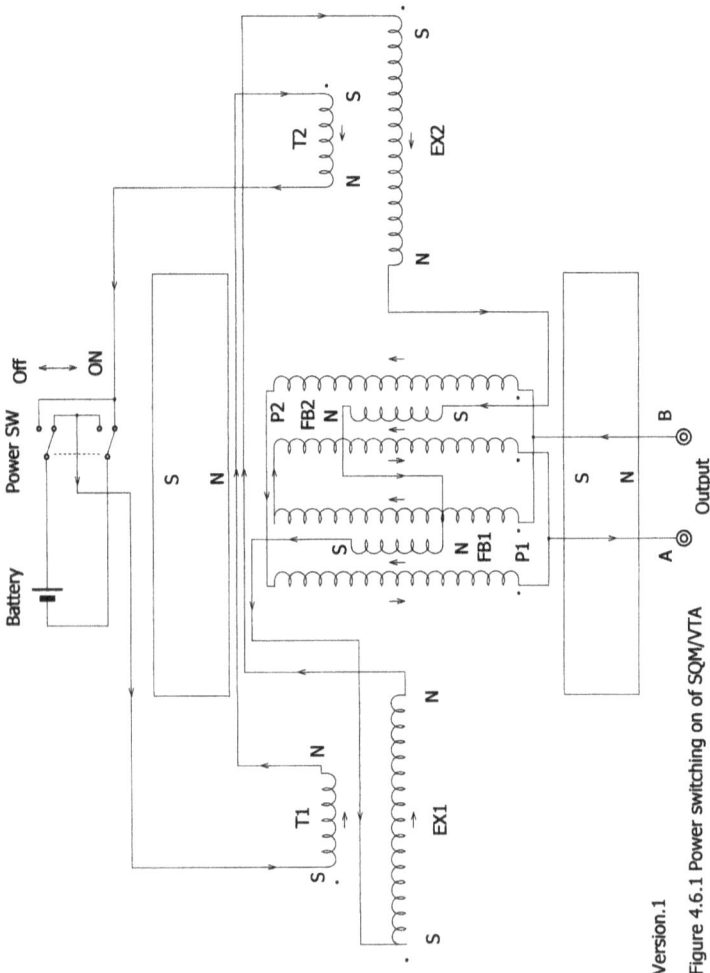

Version.1

Figure 4.6.1 Power switching on of SQM/VTA

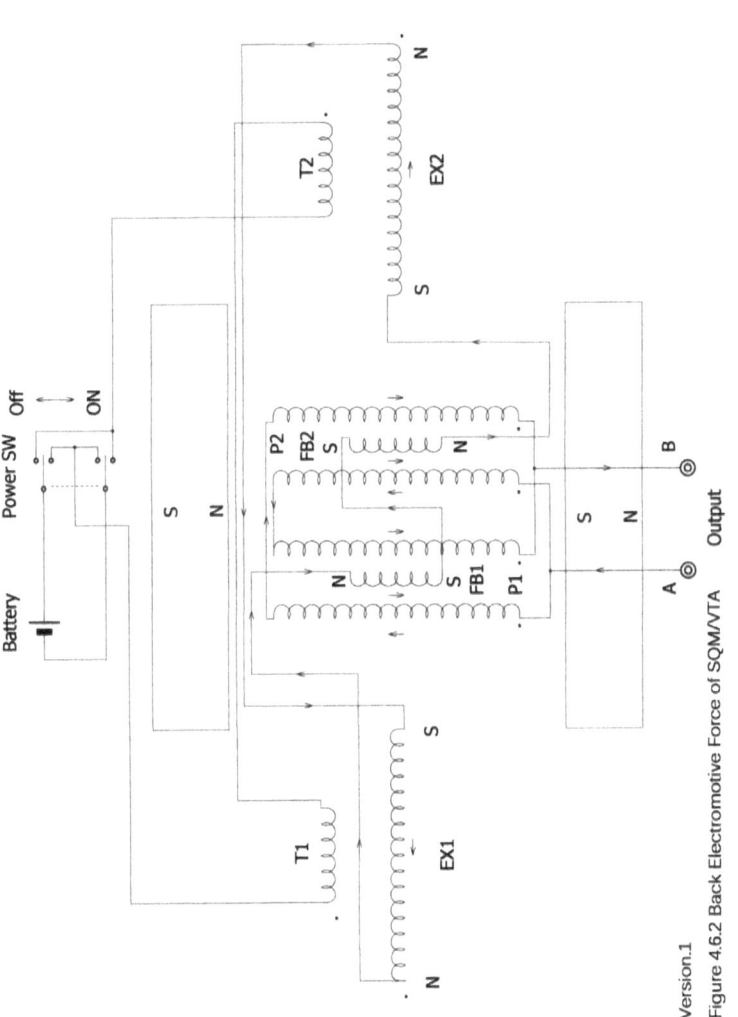

Version.1

Figure 4.6.2 Back Electromotive Force of SQM/VTA

This feedback coil FB1 is wound positive mutual inductance and the current flows from N side to S side. Feedback coil FB2 is wound negative mutual inductance and the current flows from S side to N side. FB1 and FB2 make S→N→N→S output circuit and the voltage is caused at both ends of the feedback coils FB1 and FB2.

The flux of magnetic induction of feedback coil FB1 is canceled because it becomes flux of magnetic induction and opposite direction that flows from N→S, upper part N side magnet, and lower part S side magnet to the underside and the flowing current becomes a complex current.

The flux of magnetic induction of feedback coil FB2 becomes the same direction as the flux of magnetic induction that flows from on S→N, upper part N side magnet, and lower part S side magnet to the underside while the current is increasing power. Power coil P1 / P2 resonate to "158 kHz" and increase the number of free electrons. Moreover the electron is converted into the positron so that the power coil P1 / P2 may form the Moebius ring.

3. Coil FB1 / FB2, P1 / P2 as shown in Figure 4.6.1 are placed in a negative area, trapped between upper and lower magnet. The voltage at both ends of FB1 and FB2 acts on connected magnetizing coil EX1/EX2, becoming a midair solenoid just like the T1/T2 coil. This forms a flux of magnetic induction (4.6 kHz) to the center, and the vector potential is generated outside of the coil.

The flux of magnetic induction at the center of the EX2 coil works on the right in the power coil P2 and the flux of magnetic induction at the center of the EX1 coil works on the left in the power coil P1. Flux of magnetic induction B flows from the upper part N to lower part S magnet in both N sides like a sandwich with the flux of magnetic induction.

The alternative output on the T1 / T2 coil is generated by the whirlpool of the flux of magnetic induction and the zoom of the magnetic induction in power coil P1/P2. Because it makes the continuance of the current of magnetizing coil EX1/EX2 with which the voltage at FB1 and both ends of the flow of the FB2 coil

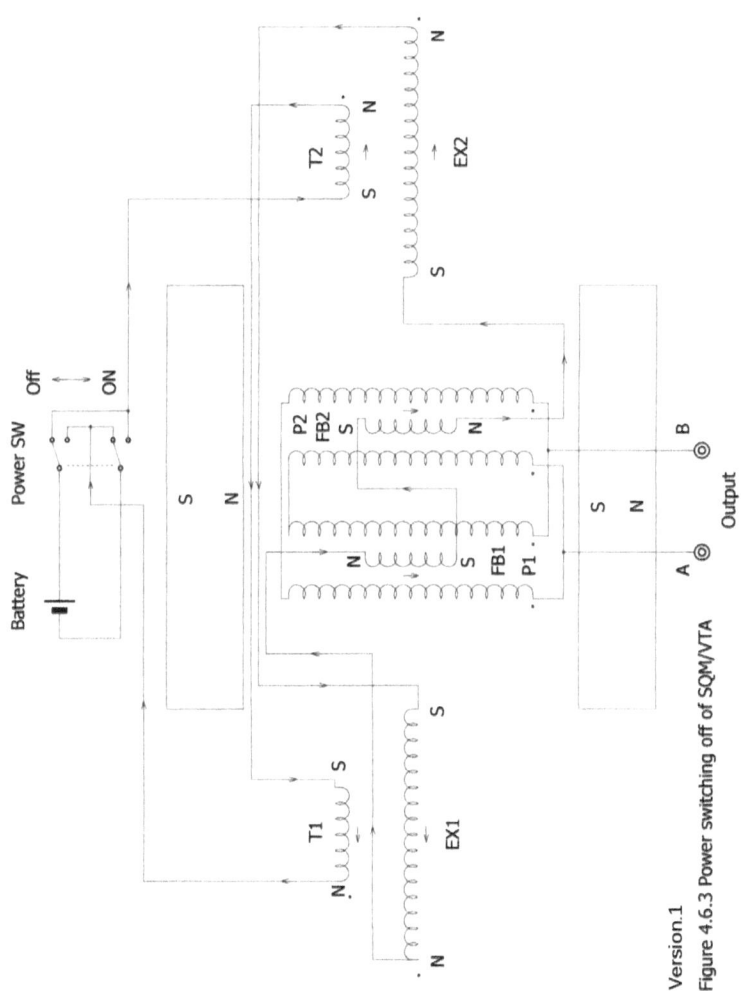

Version.1

Figure 4.6.3 Power switching off of SQM/VTA

are connected to FB1 and the FB2 coil, the transformer is united with the P1/P2 coil and their working. This makes the exchange output of power coil P1/P2 (voltage and current) steady.

EX1 and EX2 resonate to "4.6 kHz" and loosen the molecular binding.

4. Power coil P1/P2 is a deformation of bifilar coil of the Nikola Tesla invention where two coils are rolled at the same time. The electric field is formed because the current flows mutually in the current that flows to the two coils, since the opposite direction is canceled (the electric field=0). The flux of the magnetic field and the vector potential create opposing directions, canceling each other out (magnetic field=0).

When the T1/T2 coil, EX1 and the EX2 coil work, the exchange output at a half cycle is generated in power coil P1/P2. The back electromotive force is generated in power coil P1/P2 when EX1 and the EX2 coil are turned off as in Figure 4.6.2. An opposite exchange output of a half cycle is generated in power coil P1/P2.

Coil FB1 and FB2 flow in a reverse electromotive current by the back electromotive force. Coil FB1 increases in power as coil FB2 cancels the magnetic field and becomes a complex current.

5. The back electromotive force is introduced by Dr. Robert Adam's development of the Adams motor as well as Tim Harwood and John Jankowski's development of POD II.

Power coil P1/P2 becomes highly effective because it can create the exchange output of one cycle by the energy required for a half cycle.

6. The exchange output of power coil P1/P2 is controlled by the flux of magnetic induction at the center of EX1 and the EX2 coil. The current supplied to EX1 and the EX2 coil returns positive even when the overload is connected with power coil P1/P2 because it is controlled by feedback coil FB1 and FB2 that unites the transformer with power coil P1/P2.

The tornado phenomenon, which Floyd Sweet and his wife experienced, is prevented from being generated by stabilizing the

AC output (both voltage and current) from the power coil P1/P2 on this circuit.

7. If you want to stop the current as shown in Figure 4.6.3, the power switch is adjusted to the off side. The reverse current with the battery flows to the T1 / T2 coil, and the T1 / T2 coil becomes a midair solenoid and a flux of magnetic induction (35.1 kHz) opposite to turning it on to the center is formed. The vector potential opposite to turning it on outside of the T1 / T2 coil is generated.

  The flux of magnetic induction and the vector potential are expanded on the outside of the T1 / T2 coil. The supply of the flux of magnetic induction stops to P1/ P2 coils. The output of P1/P2 becomes zero by the above action and the device stops by the off control of the T1/T2 coils.

4.6.2 Discussion of Maxwell's Electromagnetic Potential Equation

1. $-E = \nabla \cdot \phi + \partial A/\partial t$
The radiation of the scalar potential from the bifilar coil P1 / P2 and the time change of the vector potential by coil T1, T2, EX1 and EX2 go to the negative area of electric field.

2. $B = \nabla \times A$
The rotation of vector potential by the outside of the coil T1, T2, EX1 and EX2 raises the magnetic field intensity of coil P1 / P2.

3. $\partial \rho/\partial t = \nabla \cdot j$
The time change of the charge from the surface of up and down the barium ferrite magnets, which are high dielectric substances, radiates current. The coil T1, T2, EX1, and EX2 change an electromagnetic charge into the current.

4.6.3 Reference

1. http://www.hyiq.org/Library/Floyd_Sweet.htm

4.7 "Tesla Coil" by Nikola Tesla

4.7.1 Analysis Result

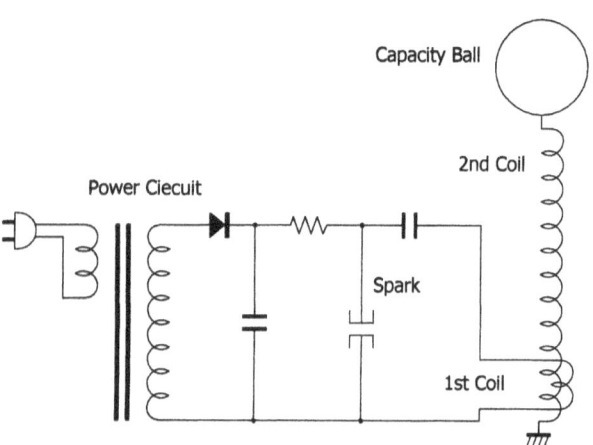

Figure 4.7 "Tesla Coil[1]"

1. Tesla Coil of Figure 4.7 shows the following characteristics:
    1. The spark discharge is placed in the primary coil side.
    2. The primary and secondary coils are resonated.
    3. The secondary coil changes an electromagnetic charge into the current.
    4. The primary and secondary coil makes a standing wave according to the progressive wave and reflected wave.
    5. The input transformer coil is wound so that the mutual inductance becomes positive.
    6. The primary coil is wound negative mutual inductance a few turns.
    7. AC power circuit is placed in the input side.
    8. Electricity is discharged with the capacity ball.

2. The spark discharge is placed in the primary coil side.
    1. The spark discharge causes the shorting of the capacitor and

coil and increases the alternating power in the wide range. It's the same as Element technology 3-5-4.
2. High voltage and the small current make the input resistance high. It's the same as Element Technology 3-7-3.

3. The primary and secondary coils are resonated.
  1. The resonance makes a strong connection between the primary and secondary coils. It is the same as Elemental Technology 3-8-5.
  2. A big Q of coils creates efficiency for the increasing power. The boost voltage is decided by a usual transformer depending on the rolling number ratio. The effect of boost voltage with Q ratio more than a usual transformer is large. It's the same as Element Technology 3-2-2.

4. The secondary coil changes an electromagnetic charge into the current. It uses the Earth similar to the magnet and collects charge (es^2) in air, creating a complex current with the capacity ball. It's the same as Element technology 3-6-2.

5. The primary and secondary coil creates a standing wave according to the progressive wave and reflected wave.
  1. It is a means of increasing power with the coil. It's the same as Element technology 3-2-2.
  2. You can see the reflected wave to be equivalent to the back electromotive force. It's the same as Element technology 3-4-4.

6. The input transformer coil is wound so that the mutual inductance becomes positive. It's equivalent to coil m3, and the same as Element technology 3-7-1.

7. The primary coil is wound negative mutual inductance a few turns. The primary coil is equivalent to the Moebius Ring. It's the same as Element technology 3-1-2.

8. AC power circuit is placed in the input side. The AC power

frequency is used as a frequency of oscillator (OSC). It's the same as Element technology 3-9-6.

9. Electricity is discharged with the capacity ball. The capacity ball discharges complex electric potential in the air. It's the same as Element technology 3-3-1.

4.7.2 Discussion of Maxwell's Electromagnetic Potential Equation

1. $-E = \nabla \cdot \phi + \partial A / \partial t$

The radiation of the scalar potential is released by the electrical discharge from the capacity ball with the secondary coil, and the time change of the vector potential by the outside of the primary coil goes to the negative area of the electric field.

2. $B = \nabla \times A$

The rotation of vector potential by the outside of the primary coil raises the magnetic field intensity.

3. $\partial \rho / \partial t = \nabla \cdot j$

The time change of the charge from the capacity ball with the secondary coil radiates the current. The secondary coil changes an electromagnetic charge into the current.

4.7.3 Summary

•Tesla Coil could be an example of having all the elemental technologies of the free energy device.

4.7.4 Reference

"Research note" 2008-0306 2005-1026 making revisions (c), Toshiki Sugiyama

Postscript

After searching for the origin of free energy, I have finally gained some clarification on "Dark Matter," "Dark Energy," "Gravity" and also on an integrated hypothesis of four powers: "Electromagnetism Power," "Gravity," "Strong Power" and "Weak Force." I could not have reached this stage without two rabbits: one who guided me, the other who connected me with the important theories.

I understand that Maxwell's electromagnetic potential equation is the basic theory of the free energy device. The majority of the elemental technology of free energy device belongs to Nikola Tesla, and his monumental achievement should be revalued as exceptional.

Also, Hans Coler's technology should be revalued as exceptional.

I welcome different opinions, should there be any. I acknowledge that I have insufficient materials and analyzed schematic diagrams. Some parts of this book might be based simply on my presumptions.

My desire is that safe and free energy devices will be multiplied soon.

Finally I would like to extend my thanks to the following people:

Professor Lisa Randall of "Warped Passages." Without this book, I might not have delved into this difficult field.

Mr. David W. Thomson III and Mr. Jim D. Bourassa, who are the scholars at Quantum Aether Dynamics Institute. I was fortunate to have purchased an autographed copy of their book, "Secrets of the Aether." I have learned a great deal from their book.

Mr. Oriharu, who is my senior of the free energy research in Japan, and who has been writing useful articles on his home page.

Index

Table of Figures   **Page**

New "Power and Energy Theory" Summary

Important Equation

1. Photon = hc = G • me • ms   c: Speed of light
   G: Newton's gravitational
        constant

2. $ms = c^2 • \lambda c/G$ h: Planck's constant
   Kc: Coulomb's constant

3. $es^2 = c^4 • \lambda c^2/(G • 16\pi^2 Kc)$ me: Electronic mass
   ms: Mass of strong power

4. $G = hc/(me • ms)$ es^2: Charge of strong power
   λc: Compton wavelength

Maxwell's electromagnetic potential equation

1. $-E = \nabla • \phi + \partial A/\partial t$ ϕ: Scalar potential
   A: Vector potential

2. $B = \nabla \times A$ E: Electric field
   B: Magnetic field

3. $\partial\rho / \partial t = \nabla • J$ ρ: Charge
   J: Current

About the Author

Makoto D. Yamane is Japanese-born, a graduate of Tottori University, engineer for Hitachi Electronics Ltd, Tottori Sanyo Electric Co., LTD, Epson Imaging Device Co., LTD, and published author.
He resides in Tottori City, Tottori Prefecture, Japan.

www.ingramcontent.com/pod-product-compliance
Lightning Source LLC
Chambersburg PA
CBHW030902180526
45163CB00004B/1667